Construction Companion to Inspecting Works

Nicholas Jamieson

RIBA Enterprises

© Nicholas Jamieson 2003

Published by RIBA Enterprises Ltd, 1–3 Dufferin Street, London EC1Y 8NA

ISBN 1 85946 098 4

Product Code: 32946

The right of the Author to be identified as the Author of this work has been asserted in accordance with the Copyright, Designs and Patents Act 1988.

All rights reserved. No part of this publication may be reproduced, stored in a retrieval system, or transmitted, in any form or by any means, electronic, mechanical, photocopying, recording or otherwise, without the prior permission of the copyright owner.

British Library Cataloguing in Publication Data
A catalogue record of this book is available from the British Library.

Publisher: Steven Cross
Commissioning Editor: Matthew Thompson
Series Editor: David Chappell
Editor: Lionel Browne
Project Editor: Anna Walters

Typeset, printed and bound by Hobbs the Printers, Hampshire

While every effort has been made to check the accuracy of the information given in this book, readers should always make their own checks. Neither the Author nor the Publisher accept any responsibility for mis-statements made in it or misunderstandings arising from it.

Contents

Foreword ix

Introduction xi

Acknowledgements xi

1 The architect's obligations to inspect 1
  1.1 What is the architect's duty of inspection? 1
  1.2 What do building contracts say? 1
  1.3 What do the standard forms of appointment say? 3
     1.3.1 The 1966 version of the RIBA Conditions of Engagement 3
     1.3.2 The 1971 version of Conditions of Engagement 5
     1.3.3 Architect's Appointment 6
     1.3.4 Standard Form of Agreement for the Appointment of an Architect 7
     1.3.5 SFA/99 10
  1.4 What are the architect's duties if nothing is specified? 13

2 How have the courts defined the architect's duties to inspect? 15
  2.1 Generally 15
  2.2 With what degree of care should the architect's duties to inspect be discharged? 16
  2.3 How often and for how long should the architect visit site? 19
  2.4 To what extent can the architect be expected to discover defects? 24
  2.5 What duties does the architect owe to the contractor in connection with defects? 27
  2.6 What duties does the architect owe to the contractor in connection with methods of working? 29
  2.7 To what extent is the architect liable for the performance of the clerk of works? 33

3 Practice management matters 39
  3.1 How can the architect contribute to the achievement of high-quality work on building sites? 39
  3.2 The architect's appointment 39

	3.3	Should the architect offer to carry out reduced or partial site inspection services?	41
	3.4	Time allowance	42
	3.5	Who should do the inspecting?	43
	3.6	Should there be a practice site inspector?	43
	3.7	Other consultants' appointments	44
	3.8	The tender documents	44
	3.9	Clauses 2.1 and 8.2.2 of JCT 98	46
	3.10	Is a clerk of works or site architect required?	47
	3.11	Finding a clerk of works	49
4	**As work is about to start**	51
	4.1	Importance of pre-planning and prioritising	51
	4.2	Categories of inspection	52
	4.3	Importance of the contract documents	53
	4.4	Time monitoring	54
	4.5	Advising the client	54
	4.6	Briefing site inspectors	54
	4.7	Initial project team meeting	55
	4.8	The contractor's programme	56
	4.9	First meeting with the site agent	57
5	**While work is in progress**	59
	5.1	The timing of inspections	59
	5.2	Use of checklists	59
	5.3	Priorities for inspection	61
	5.4	Once on site	61
	5.5	Contractual provisions	63
	5.6	Storage and protection	64
	5.7	Inspecting work off site	65
	5.8	The contractor's quality control procedures	66
	5.9	Contractor's records	67
	5.10	What should the architect do if defective work is found?	68
	5.11	Health and safety	71
	5.12	Site security	72
	5.13	Welfare facilities	72
	5.14	Site tidiness	73
	5.15	Monitoring progress	73
	5.16	Daywork sheets	74
	5.17	The clerk of works and other site inspectors	74
	5.18	Other consultants	75

		5.19 Inspection records	76
		5.20 Completion	77
		5.21 At the end of the contract	79
6	**Practical matters**		81
	6.1	What should the architect wear on site?	81
	6.2	What else should the architect take on site?	81
	6.3	How should the architect conduct himself or herself on site?	82
	6.4	Safety on site	85
	6.5	Personal possessions	85
	6.6	What should the architect do if contractor's personnel behave in an uncooperative, obstructive or intimidating manner?	85
	6.7	Building up knowledge	86
	6.8	How should the architect deal with queries raised by the contractor on site?	86
	6.9	Can the architect do too much inspection?	87
	6.10	What should the architect do if the contractor tries to persuade the architect to accept defective work?	88
	6.11	What should the architect do if the contractor suggests replacing a specified material or work process with an alternative?	88
	6.12	What should the architect do if the contractor is slow in making good defective work?	89
	6.13	What should the architect do if suspecting the contractor has covered up incomplete or defective work?	89
7	**Inspecting the work**		91
	7.1	Inspecting generally	91
	7.2	Preliminaries	91
	7.3	Demolition and site stripping	92
	7.4	Setting out	92
	7.5	Substructure, below-ground drainage, and structural concrete	93
		7.5.1 Generally	93
		7.5.2 Piling	93
		7.5.3 Excavations	93
		7.5.4 Below-ground drainage	94
		7.5.5 Hardcore	96
		7.5.6 Damp-proof membranes and insulation below slabs	96
		7.5.7 Reinforcement	97
		7.5.8 Formwork	97
		7.5.9 Concreting	98
		7.5.10 Basements	101

	7.5.11 Precast concrete floors	101
	7.5.12 Damp-proof membranes and insulation above concrete floors	102
	7.5.13 Screeds	102
7.6	Structural steelwork	103
	7.6.1 Generally	103
7.7	Timber structure	105
	7.7.1 Generally	105
	7.7.2 Suspended timber floors and ceilings	105
	7.7.3 Timber framing and studwork	106
	7.7.4 Roofs	107
	7.7.5 Stairs	108
7.8	Masonry	108
	7.8.1 Generally	108
	7.8.2 Materials and storage	109
	7.8.3 Workmanship generally	110
	7.8.4 Cavity work	112
	7.8.5 Protection of new brickwork	113
7.9	Roof finishes	114
	7.9.1 Generally	114
	7.9.2 Tiling and slating	115
	7.9.3 Profiled metal or fibre reinforced cement roofing	116
	7.9.4 Fully supported metal roofing	117
	7.9.5 Built-up and single-ply roofing	117
	7.9.6 Asphalting	118
7.10	External wall finishes	119
	7.10.1 Render	119
	7.10.2 Tiles	120
	7.10.3 Claddings generally	120
7.11	Windows and doors	121
	7.11.1 Components generally	121
	7.11.2 Frames and leaves	121
	7.11.3 Glazing	122
7.12	Services	123
	7.12.1 Generally	123
	7.12.2 Electrical services	124
	7.12.3 External above-ground drainage	124
	7.12.4 Internal above-ground drainage, and heating and hot and cold water services	125
7.13	Internal finishes	127
	7.13.1 Plasterboarding	127
	7.13.2 Plastering	128
	7.13.3 Wall and floor tiling	129
	7.13.4 Timber and sheet flooring	130

7.14		Built-in furniture and fittings	130
7.15		Painting and decorating	131
	7.15.1	Generally	131
	7.15.2	External painting	131
	7.15.3	Internal painting	132
	7.15.4	Wall coverings	132
7.16		External works	132
7.17		Practical completion	133

Bibliography 135

Index 137

Foreword

This is the sixth of a series of guides being produced by RIBA Publications under the general heading of *Construction Companions*. They are intended to be compact and accessible guides written in plain language, but each one authoritative in its own subject.

Inspecting work on site is a tricky area. Few schools of architecture appear to think it a worthwhile subject and yet it is perfectly obvious that the ability to inspect work in progress is absolutely fundamental to the business of getting a building built. Most architects learn how to inspect by a mixture of trial and error, and the odd spot of advice from someone in the office. The result is that the first time an architect visits site alone can be a nightmare and is seldom useful. Despite the fact that it is the contractor's duty to construct a building in accordance with the contract, it is common for architects to be accused of failure to inspect properly. Architects cannot inspect every part of a building and check all the materials. The question is: which parts of a building should be inspected and how should it be accomplished?

So far as I am aware, this is the only book to cover this topic in detail. It sets out the architect's duty to inspect as indicated in various terms of engagement, what that really means in law, and how to manage the process efficiently to guard against allegations of negligence. It concludes by explaining the practical aspects of inspecting work on site. It is written in a clear, readable style, even when dealing with the notoriously difficult legal principles.

It is a very fine exposition of the subject by an experienced architect. I wish a book of this quality had been available when I started practice many years ago.

David Chappell BA(Hons Arch) MA(Arch) MA(Law) PhD RIBA
Series Editor

David Chappell has worked as an architect in the public and private sectors and is currently Director of Chappell-Marshall Limited, a building and civil engineering contracts consultancy. He is Professor of Architectural Practice and Management Research at The Queen's University of Belfast and recently appointed Visiting Professor in Practice Management and Law at UCE Birmingham. He is a frequent speaker and author of many books for the construction industry.

Introduction

If a client employs an architect to certify payments to a contractor, the architect must inspect work in progress. This book provides guidance on how to carry out such inspections.

The book is divided into three parts. The first part (Chapters 1 and 2) describes the legal context in which inspections are carried out. The second part (Chapters 3–6) gives general practical advice. The final part (Chapter 7) provides tips on what to look out for when inspecting the work of specific trades.

The guidance in the book can be summarised by saying that the architect should check everything that is important, and keep records showing that the checks have been carried out. The rest of the book is an elaboration of this essential principle.

Acknowledgements

Thank you to David Clarke, Mark Sutcliffe, Barry Copeland, John Potter, Paul Hyett and Barbara Weiss for giving me the opportunity to gain the practical experience upon which much of this book is based; to John Edwards for comments on structural matters; to David Chappell for inspiration and legal advice; to Matthew Thompson for his enthusiasm and guidance; to everyone at Barbara Weiss Architects for their moral and practical support; and to Lizzy for all her comments, patience, and encouragement.

1 The architect's obligations to inspect

1.1 What is the architect's duty of inspection?

Vital to successful practice is a clear understanding of the architect's responsibilities, and the extent of the architect's liabilities in connection with building work being carried out on site.

The architect's duty to inspect is defined by the terms, both explicit and implied, of the architect's appointment. There is, however, a wide range of different services for which an architect may be appointed, and a number of different standard forms that may be used.

The RIBA alone publishes three main appointment documents, with various amendments and guidance to allow for the particular requirements of design and build contracts, historic buildings, and community architecture. Other documents are published by the Association of Consultant Architects, and by individual clients such as the National Health Service. A standard set of appointment documents published by the Joint Contracts Tribunal has been on the horizon for a number of years. Each document says something different about inspection.

In addition, a practical understanding of the architect's responsibilities and liabilities in relation to work in progress cannot be gained simply by reading even the most elaborate of the forms referred to above – it is necessary for the practising architect to look also at the architect's duties and powers under building contracts, and at how the architect's duties have been defined by the courts.

1.2 What do building contracts say?

Architects often think of their primary function as being the designing of buildings. Indeed, under some contemporary methods of procurement the architect is required to do no more than prepare design and production information. However, historically, just as much as to design buildings, the profession was brought into being to give its clients confidence that:

- the quality of the work being carried out by a client's building contractor complied with the standards agreed under the contract between the client and the contractor
- the client was not being misled into paying for work that did not comply with the agreed standards.

The need for such a role was born from the divided interests of builders and those who employed them.

The tensions between the concerns of employer and contractor continue to be reflected in the wording of most building contracts used today, and despite changing procurement trends many contracts continue to rely for the resolution of such tensions on the functions of an independent architect.

Under such building contracts an architect is required to issue certificates for the value of work properly executed in accordance with the contract. By implication, before such a certificate can be issued the architect must be satisfied that the work to which the certificate relates has actually been carried out and has been carried out correctly. The architect does so by visiting the site and inspecting the work as it is being built.

The procedure whereby the architect inspects the contractor's work and then certifies payment for work correctly done – and only for work correctly done – is fundamental to the operation of the contract. It is also fundamental to the protection of the employer's interests, and is one of the principal motives for the architect's retention by the architect's client during the construction stages of the project.

It follows that, if an architect agrees to act as architect under the terms of such a contract, it becomes part of his or her job to see that if work is not done properly the contractor does not get paid for it – even though the architect's appointment may make no specific reference to such a duty.

At the same time, it is important that architects, employers and contractors understand that under the building contract the architect needs to check work only for the purpose of certification; the architect does not owe a duty to the contractor to spot or report defects. (As far as the contractor is concerned it is entirely his own responsibility to check work in progress as necessary to ensure that it is completed in accordance with the contract; he is not permitted to rely on the architect.)

It is also important that architects and their clients understand that the ability of the architect to protect the employer against defective work by the contractor is limited. Typically, the building contract will give the architect power only:

- to refuse to certify payments for defective work
- to order the removal of defective work from site
- to issue instructions requiring that defective work be remedied within a given period
- ultimately to determine the contractor's employment.

Contrary to some clients' expectations, building contracts do not give the architect power to take the contractor by the scruff of the neck and force him to carry out the works in accordance with the contract.

In simple terms, the main function of the inspecting architect is to see that the client gets value for money: the architect cannot be blamed for the failures of the contractor, but can be blamed for negligently certifying payment for work that is not done properly.

1.3   What do the standard forms of appointment say?

The extent to which the architect, before certifying payment, should inspect work in progress is determined by the wording of the architect's appointment, and its interpretation by the courts.

To understand architects' inspection duties as described by current standard appointment documents it is helpful to look back to previous documents, published when the appointment of an architect seems to have been simpler, and to trace subsequent developments.

1.3.1   The 1966 version of the RIBA Conditions of Engagement

In 1966 the RIBA published the first version of Conditions of Engagement, which states that, irrespective of the nature of the project:

> *1.16 The architect shall give such periodic supervision and inspection as may be necessary to ensure the works are being executed in general accordance with the contract; constant supervision does not form part of his normal duties.*

> *1.17 Where the employment of a resident architect for constant supervision is agreed, he will be employed by the architect . . .*

> *1.18 Where frequent or constant inspection is required a Clerk of Works shall be employed. He shall be nominated or approved by the architect, and be under the architect's direction or control. He shall be appointed and paid by the client or alternatively may be employed by the architect . . .*

The document continues, in clause 2.50, to include in the services to be provided by the architect:

> *. . . Periodic site supervision, issuing certificates and other administrative duties under the building contract . . .*

The first point to note is the distinction in clause 1.16 between 'periodic' and 'constant' supervision and inspection. The distinction goes to the root of what is normally to be expected of the architect in relation to work in progress on site.

All forms of appointment, when setting out the inspection duties of the architect, attempt to reconcile:

- the interests of the client
- the interests of the architect
- the requirements of the building contract.

It might be expected that clients would demand of their architects continuous presence on site, that architects maintain a constant watch over the shoulder of every one of the contractor's operatives, that architects check the length and gauge of every screw and nail being fixed – thereby giving clients the greatest confidence that they are getting that for which they will be paying. However, a need for the architect to perform such a service would imply that the contractor is either totally incompetent or totally untrustworthy, neither of which should be the case if the contractor has been correctly chosen. Therefore, payment of the fees that an architect would charge for performing such services is not normally justified, and agreements between architects and clients have been designed to give the client as much peace of mind as can be afforded without wasting too much money paying the architect to double-check construction work.

The primary purpose of the distinction in clause 1.16 between periodic and continuous supervision and inspection is therefore to make sure that clients do not automatically assume that the architect will be following the contractor's every move on site, but that the architect can be expected to inspect the works as often as is necessary to ensure that they are 'in general' being carried out as described by the contract documents. If the client requires more frequent inspection or supervision, clauses 1.17 and 1.18 provide for the appointment of specialist site staff.

Of further interest is the use of the terms 'supervision' and 'inspection', and the subtle differentiation that the document makes between them. Insofar as 'periodic supervision and inspection' are required the architect is to be responsible for both; but if 'constant supervision' is required a resident architect is to be employed, and if 'constant inspection' is required a clerk of works is to be employed. The allocation of supervision to the resident architect, and of inspection to the clerk of works, implies that the responsibilities of supervision are the more onerous. The responsibilities for the architect were apparently found too onerous: after initial publication of the document it was judged that the term 'supervision' implies duties that, under the terms of conventional building contracts, should properly be the responsibility of the contractor rather than of the architect – and the term was omitted from all subsequent standard forms for the appointment of an architect. (Members of the judiciary and other legal professions have, however, continued habitually to refer to 'supervision' rather 'inspection'.)

A further point to note is that clause 1.16 leaves the architect with the duty to 'ensure' the works are being carried out in accordance with the contract (albeit only 'in general'). The wording appears to require the architect, if necessary, somehow to force the contractor to carry out the works in accordance with the contract documents – but does not make clear how the architect is effectively empowered to do so. As the architect is in practice incapable of forcing the contractor to do anything, the word 'ensure' was omitted from the second edition of the document.[1]

1.3.2 The 1971 version of Conditions of Engagement

The next version of Conditions of Engagement was published in 1971. It states:

> 1.33 The architect shall ... make such periodic visits to the site as he considers necessary to inspect generally the progress and quality of the work and to determine in general if the work is proceeding in accordance with the contract documents.

> 1.34 The architect shall not be responsible for the contractor's operational methods, techniques, sequences or procedures, nor for safety precautions in connection with the work, nor shall he be responsible for any failure by the contractor to carry out and complete the work in accordance with the terms of the building contract between the client and the contractor.

It continues:

> 1.60 During his on-site inspections made in accordance with Clause 1.33 the architect shall endeavour to guard the client against defects and deficiencies in the work of the contractor, but shall not be required to make exhaustive or continuous inspections to check the quality or quantity of the work.

> 1.61 Where frequent or constant inspection is required a clerk or clerks of works should be employed. He shall be nominated or approved by the architect and be under the architect's direction and control. He may be appointed and paid by the client or employed by the architect.

> 1.62 Where the need for frequent or constant on-site inspection by the architect is agreed to be necessary, a resident architect shall be appointed by the architect.

The document continues, in clause 2.11, to include in the services to be provided by the architect:

> ... *Making periodic visits to the site as described in clause 1.33; issuing certificates and other administrative duties under the contract* ...

Clauses 1.33, 1.34 and 1.60 of the 1971 version cover the same ground as clause 1.16 of the 1966 version, and clause 2.11 the same as the previous clause 2.50. However, the new clauses are extended with careful rewording and qualification aimed at limiting the architect's liability in respect of failures by the contractor.

In particular, the word 'supervision' is completely excised, and the wording of clause 1.16 of the 1966 version requiring the architect to 'ensure' the works are being executed in accordance with the contract is replaced in clause 1.33 of the 1971 version by the requirement for the architect merely to 'determine if' the work is being carried out as it should be, and in clause 1.60 by the requirement for the architect to do no more than 'endeavour to guard' against any shortcomings of the contractor. Clause 1.34 emphasises the latter point by listing specific exclusions.

Clause 1.61 of the 1971 version is practically identical to clause 1.18 of the previous version, and clause 1.62 of the 1971 version corresponds to clause 1.17 of the earlier version, but is rewritten to avoid the use of the term 'supervision'.

Clause 1.33, for the first time in an RIBA form of appointment, explicitly requires the architect to inspect progress as well as quality.

In general, the changes incorporated within the 1971 version of Conditions of Engagement serve to align the requirements of the architect's appointment more accurately with the architect's duties under the building contract, placing responsibility for the contractor's performance where it properly belongs, firmly with the contractor, while leaving responsibility for carrying out inspections for the purpose of certifying payments with the architect.

1.3.3 Architect's Appointment

In 1982 the RIBA superseded Conditions of Engagement with Architect's Appointment. In 'Part 1 Architect's Services', which describes services normally to be provided, the architect is required to:

> *1.21 Administer the terms of the building contract during operations on site.*

and

1.22 Visit the site as appropriate to inspect generally the progress and quality of the work.

In 'Part 3 Conditions of Appointment', the document states that when employed in connection with the construction stages of a project:

3.10 ... the architect will visit the site at intervals appropriate to the stage of construction to inspect the progress and quality of the works and to determine that they are being executed generally in accordance with the contract documents. The architect will not be required to make frequent or constant inspections.

3.11 Where frequent or constant inspection is required a clerk or clerks of works will be employed. They will be employed either by the client or by the architect and will in either event be under the architect's direction and control.

3.12 Where frequent or constant inspection by the architect is agreed to be necessary, a resident architect may be appointed by the architect on part or full-time basis.

The wording of Architect's Appointment is a more succinct equivalent to that of the later version of Conditions of Engagement.

1.3.4 Standard Form of Agreement for the Appointment of an Architect

Architect's Appointment and Conditions of Engagement were products of a time when it would be assumed that, whatever the project, the architect would act, simply, as architect. It was common knowledge, even among those unfamiliar with the construction industry, that anyone playing the role of architect could be expected to act as principal designer, to be leader of the project team, and generally to bear overall responsibility for the project. So much was the role taken for granted that within appointment documents it was necessary only very briefly to define the services to be performed by the architect. However, by 1992, when the RIBA replaced Architect's Appointment with Standard Form of Agreement for the Appointment of an Architect (SFA/92), the old idea of the architect as a trusted professional advisor had been replaced by the architect reconceived as a provider of services.[2]

SFA/92 reflected the shift in political values by, in 'Schedule Two', presenting a catalogue of potential services from which clients could pick and choose the particular services they wanted. On offer for customers interested in inspection services were, under the heading 'K-L Operations on Site and Completion', the following:

04 Generally inspect materials delivered to site

05 As appropriate conduct sample taking and carrying out tests of materials, components, techniques and workmanship and examine the conduct and results of such tests whether on or off site

06 As appropriate instruct the opening up of completed work to determine that it is generally in accordance with the Contract Documents

07 As appropriate visit the sites of the extraction and fabrication and assembly of materials and components to inspect such materials and workmanship before delivery to site

08 At intervals appropriate to the stage of construction visit the Works to inspect the progress and quality of the Works and to determine that they are being executed generally in accordance with the Contract Documents

09 Direct and control the activities of Site Staff...

14 Monitor the progress of the Works against the contractor's programme and report to the Client.

Any of the above services chosen were to be carried out subject to conditions, set out under the heading 'Conditions of Appointment', as follows:

3.1.1 The Architect shall in providing the Services specified in stages K and L of Schedule Two make such visits to the Works as the Architect at the date of the Appointment reasonably expected to be necessary. The architect shall confirm such expectation in writing.

3.1.2 The Architect shall, on its becoming apparent that the expectation of the visits to the Works needs to be varied, inform the Client in writing of his recommendations and any consequential variation in fees.

3.1.3 The Architect shall, where the Client requires more frequent visits to the Works than that specified by the Architect in condition 3.1.1, inform the Client of any consequential variation in fees. The Architect shall confirm in writing any agreement reached...

> *3.3.1 The Architect shall recommend the appointment of Site Staff to the Client if in his opinion such appointments are necessary to provide the Services specified in K-L 04-08 of Schedule Two.*
>
> *3.3.2 The Architect shall confirm in writing to the Client the Site Staff to be appointed, their disciplines, the expected duration of their employment, the party to appoint them and the party to pay, and the method of recovery of payment to them.*
>
> *3.3.3 All Site Staff shall be under the direction and control of the Architect.*

Under the heading 'Definitions' the document defines 'Site Staff' as:

> *Staff appointed by either the Architect or the Client to provide inspection of the Works on behalf of the Client.*

A significant change from previous forms is that clause 08 of stages K-L in 'Schedule Two' does not explicitly exclude constant, exhaustive, continuous or frequent inspection, albeit the architect's duties under the clause are limited by the use of the word 'generally'.

In addition clauses 3.1.1–3.1.3 of the 'Conditions' require the architect at the time of appointment to estimate the time it will take to carry out the K-L stage services the architect has been asked to provide, and to confirm the estimate in writing to the client, after which the time to be spent may be varied upon either the recommendation of the architect or the request of the client – and fees adjusted accordingly. Reasons for such variations could include the client leading the architect at the time of appointment to believe that an experienced contractor would be used, and then at tender stage insisting on the employment of an inexperienced contractor whose work would require more inspection; or an unexpected increase in the complexity of the project.

A further change from previous forms is that the collective term 'site staff' is used in place of separate references to clerk of works and site architect. It is unclear whether the change was motivated by a desire to reduce wording, or to allow for the accommodation of other types of site inspector.

In 1995, as an alternative to SFA/92 for use in small to medium-sized projects, the RIBA published Conditions of Engagement for the Appointment of an Architect (CE/95). The descriptions of inspection services, and the conditions under which they were to be performed are identical to those of SFA/92.

The greatest benefit of SFA/92 and CE/95 is that they provide the fullest description, published so far in an RIBA standard form of appointment, of the potential duties of an architect employed to provide inspection services.

1.3.5 SFA/99

In contrast to SFA/92 its replacement, SFA/99, says less about site inspection duties than any of its predecessors. In fact, nowhere in SFA/99 is there stated an express requirement for the architect to visit the site for the purpose of carrying out inspection duties. Instead there are a number of more obtuse references to site-related activities, as follows.

'Schedule 2: Services' requires the architect to:

> *Make visits to the Works in accordance with clause 2.8 [of the Conditions of Engagement].*

Clause 2.8 of the 'Conditions of Engagement' requires that:

> *The Architect shall in providing the Services make such visits to the Works as the Architect at the date of the appointment reasonably expected to be necessary.*

'Schedule 2: Services' also requires the client and the architect to choose one or more of the following roles in which the architect is to act:

- designer
- design leader
- lead consultant during pre-construction work stages
- lead consultant during construction work stages.

The roles are defined in the 'Services Supplement', which is divided into two parts. The first part is headed 'Architect's Design Services' and comprises a very abbreviated version of the list of 'Services Specific to Building Projects' provided in 'Schedule Two' of SFA/92. The second part, headed 'Architect's Management Services', defines the roles referred to above, and states that when acting as 'Lead Consultant' and 'Contract Administrator' the authority and responsibility of the architect includes:

> *administering the building contract, including:*
> - *monitoring the progress of the Works against the Contractor's programme;*
> - *issuing information, instructions, etc;*

- *preparing and certifying valuations of work carried out or completed and preparing financial reports for the Client; or*
- *certifying valuations of work prepared by others, and presenting financial reports prepared by others to the Client*
- *collating record information including the Health and Safety File*

and

co-ordinating and monitoring the work of Consultants and Site Inspectors, if any, to the extent required for the administration of the building contract, including:
- *receiving reports from such Consultants and Site Inspectors to enable decisions to be made in respect of the administration of the building contract*
- *consulting any Consultant or other person whose design or specification may be affected by a Client instruction relating to the constriction contract, obtaining any information required and issuing any necessary instructions to the Contractor;*
- *managing change control procedures and making or obtaining decisions as necessary for cost control during the construction period;*
- *providing information obtained during the administration of the building contract to the Consultants and Site Inspectors.*

'Schedule 4: Other Appointments' provides for details of a 'Site Inspector/Clerk of Works' to be entered if applicable.

The 'Conditions of Engagement', under the heading 'Definitions', define 'Site Inspectors' as:

Clerks of Works or others appointed by the Client to perform inspection services in connection with the construction of the Works.

Clause 2.5 of the 'Conditions of Engagement' requires that:

The Architect shall advise the Client on the appointment of full- or part-time Site Inspectors other than those named in Schedule 4, under separate agreements where the Architect considers that the execution of the Works warrants such appointment.

Clause 3.10 of the 'Conditions of Engagement' requires that:

Where it is agreed Site Inspectors shall be appointed they shall be under the direction of the Lead Consultant and the Client shall appoint and pay them under separate agreements and shall confirm in writing to the Architect the services to be performed, their disciplines and the expected duration of their employment.

There is no further reference to inspection services.

The differences between SFA/99 and SFA/92 are remarkable. The drafting of SFA/99 was aimed at producing an appointment document that would accord with an increasing tendency for the management functions of the architect to be separated from the design functions. Under a widening range of procurement regimes architects were finding themselves with tasks more and more narrowly defined than in the past. Assumptions could no longer be made about the inspection services – if any – that a client would require of an architect. However, whereas SFA/92 helpfully lists a wide range of services from which a client can pick and choose, SFA/99 is adaptable only insofar as it is vague. Inexperienced clients reading 'Schedule Two' of SFA/92, under the heading 'K-L Operations on Site and Completion', would find themselves informed by a fairly comprehensive summary of the inspection services that may potentially be provided by an architect. The same clients having read SFA/99 would remain ignorant of:

- the potential inspection services that the architect can provide
- the purpose of the inspection services
- the basis on which the length and frequency of the architect's visits to site should be calculated
- the limitations of the architect's liability in connection with the inspection services on offer
- how any of the above is to differ according to the management role the architect is playing.

Presumably the architect can – and should – explain such matters to the client while the appointment is being negotiated, and record the substance of any associated agreement by somehow altering, amending, or adding to the standard form, perhaps by a letter. If the architect neglects to do so, explanations may be left to the courts.

Finally, it is of interest to note that in an attempt to limit the architect's vicarious liability for negligence of 'Site Inspectors' the form, unlike all previous forms, provides no alternative to their appointment by the client. Furthermore, and for no clear reason, the form, unlike SFA/92, refers to the clerk of works but, like SFA/92, does not acknowledge the site architect – perhaps again to limit the architect's potential liability.

CE/99, the replacement for CE/95, contains almost identical wording to SFA/99, except for some rearrangement, and the omission of the definitions of the architect's management roles (although the form still requires the role or roles to be chosen).

1.4 What are the architect's duties if nothing is specified?

The duties that will be imposed upon an architect who undertakes to carry out site inspection services without making clear the nature of the services will of course depend upon the circumstances of the case. Needless to say it is not desirable for an architect to be found in such a position. However, in the past, when such cases have come to court, judges have sometimes turned for guidance to the edition of the RIBA *Architect's Job Book* current at the time of the architect's appointment.[3]

Notes

1 The question was examined in the case of *Department of National Heritage* v *Steensen Varming Mulcahy & Others* (1998) CILL 1422. The terms of its appointment required an electrical services consultant to 'make visits of inspection to ensure that the Works are being properly supervised and executed in accordance with the design and specifications'.
It was alleged that the word 'ensure' should be strictly interpreted, and that therefore any electrical defect was in itself evidence of negligence on the part of the consultant. Judge Bowsher QC, an Official Referee, disagreed – on the grounds that the consultant had been given neither power of enforcement over the electrical contractor, nor payment for taking responsibility for the electrical contractor's performance. In the context of its appointment the judge said the consultant's duties were limited 'to taking steps which would discourage bad work and if possible discover it after it had been done'.
2 By the same doctrine patrons of architecture, although still referred to as clients, were presumably to suffer the same fate as railway passengers, and become mere customers.
3 Judges have also used such guidance even when descriptions of site inspection services have been specified.

2 How have the courts defined the architect's duties to inspect?

2.1 Generally

A certain amount of care must be taken when reading case law as it will always deal with the particular circumstances of each case, not least of which will be the terms of the architect's appointment. However, it is possible to derive some general principles that can, with some confidence, be applied to an architect's duties to inspect under the standard forms of appointment.

In *Sutcliffe* v *Chippendale & Edmondson* (1971)[1] Judge William Stabb QC said:

> *It can be said that when a person engages an architect ... he is entitled to expect that the architect ... will do all that is reasonably within his power to ensure that the work is properly and expeditiously carried out, so as to achieve the end result as contemplated by the contract. In particular the building owner is entitled to expect his architect so to administer the contract and supervise the work, as to ensure, so far as is reasonably possible, that the quality of work matches up to the standard contemplated.*

It is essential to keep in mind when considering the judge's words that:

- the architect's duty to inspect is owed to the building owner alone, and not to the contractor
- the standard against which the quality of the work is to be assessed is the standard set out in the contract documents as agreed between the employer and the contractor – and is not to be subject to the architect's discretion, except where the contract expressly provides otherwise.

It is also worth remarking on the judge's use of the word 'supervision', rather than 'inspection'. The choice of words is in accordance with both the version of RIBA Conditions of Engagement and common parlance as they were at the time of the judgment. Although nowadays architects are encouraged to use the word 'inspection', the word 'supervision' is still occasionally heard. It is interesting to note Mr Recorder Coles' comments on the usage made in the later case of *Brown and Brown* v *Gilbert-Scott and Payne* (1992).[2] Referring to the case pleadings the judge, an Official Referee, said:

> *Some argument was addressed to the question whether the use of the word 'supervision' here added anything to the [architect's] obligation – i.e., above and beyond what was involved in the obligation to inspect the works as they progressed.*

> *In my judgement the [architect] had a duty to inspect the works of the [contractor] and that the use of the word 'supervision' does not enlarge his duty in any way.*

In an even more recent case, *Consarc Design Ltd* v *Hutch Investments Ltd* (1999),[3] in which it was alleged architects employed under SFA/92 failed to spot defective preparation for a screed, Judge Peter Bowsher QC said:

> *The older forms of contract required the architect to 'supervise'. The more recent contracts, including the contract in this case, require the architect to 'visit the Works to inspect the progress and quality of the Works'. It seems to me that inspection is a lesser responsibility than supervision.*

In view of the above, and despite the dicta in *Brown and Brown*, it appears that an architect using the word 'supervision' in relation to the architect's duties under his or her appointment would be unwise.

Before the judge's statement in *Sutcliffe* v *Chippendale & Edmondson* can usefully be applied in practice, it is necessary to understand how the limits of what is 'reasonably possible' are to be defined. Such an understanding can be gained by looking at how the courts have answered a number of component questions.

2.2 With what degree of care should the architect's duties to inspect be discharged?

In general, the courts have found that the degree of care required depends upon the circumstances of the case.

Some of the circumstances that it might be necessary to take into account are indicated by the case of *The Kensington Chelsea and Westminster Area Health Authority* v *Wettern Composites and Others* (1984).[4] The health authority employed a firm of architects to design an extension to a hospital. The design included artificial stone mullions, arranged in a series of columns each of five mullions one above the other, as part of the external cladding. The mullions were each about 4 m long, and weighed about 340 kg.

The mullions were to be supported vertically by either concrete corbels or metal angles, which were to fit into a recess at the back of each mullion. The mullions were to be supported horizontally by cramps, at one end of which were to be dowels fitted into the mullions at horizontal joints, and the other ends of which were to be free to slide up and down within abbey slots cast into the structural frame. The joints between mullions were to be filled with sealant to allow relative movement.

During construction a clerk of works was employed by the health authority.

Fixing of the mullions started in November 1964. Shortly afterwards the architect wrote to the contractor complaining of defects in the manufacture of the mullions, and of unacceptable attempts at rectification. Instances of other defects were recorded during construction. Fixing of the mullions was completed in July 1965.

In 1976 the health authority discovered cracking in one of the mullions. Concern was raised, the defect was monitored, and continuing movement was found. After further investigations it was concluded that the mullion installation was unsafe. In 1980 all the mullions were removed and replaced.[5]

As any claim in contract was barred by the Limitation Acts, the health authority sued the architect and the engineer in the tort of negligence in respect of their duties to 'supervise' the erection of the mullions.

In considering the skill and care with which the architect should have carried out the duty to inspect, Judge David Smout QC, Official Referee, said:

> *The ordinary skill and care must of necessity be measured with regard to the special circumstances that need to be taken into account. Amongst the special circumstances that arise in this case are the following:*
>
> - *That the architects had been alerted by 27 November 1964, to the poor workmanship and to some lack of frankness on the part of the subcontractors that should have put the architects on their guard. They became aware of further inadequate workmanship on 9 December 1964. I echo the words of Judge Stabb again in Sutcliffe v Chippendale & Edmondson...*
>
> *'I think that the degree of supervision required of an architect must be governed to some extent by his confidence in the contractor. If and when something occurs which should indicate to him a lack of competence in the contractor, then, in the interests of his employer, the standard of his supervision should be higher. No one suggests that the architect is required to tell a contractor how his work is to be done, nor is the architect responsible for the manner in which the contractor does his work. What his supervisory duty does require of him is to follow the progress of the work and to take steps to see that those works comply with the general requirements of the contract in specification and quality. If he should fail to exercise his professional care and skill in this respect, he would be liable to his employer for any damage attributable to that failure.'*
>
> - *That much of the work of fixing up the mullions would be speedily covered up in the course of erection and as such called for closer supervision than would otherwise be the case ...*

- *That the burden of supervision is the greater where poor workmanship can result in physical danger. The size and weight and position of the mullions are relevant factors.*
- *That a clerk of works was engaged to attend site full-time for the purpose of providing constant supervision of the building work so far as that was practicable.*

In conclusion, when determining the degree of care with which the architect should inspect, he or she should take into account such matters as:

- the competence of the contractor and subcontractors
- how quickly important work will be covered up
- the implications of the failure of work
- the presence of site inspectors.

The competence of the contractor was held to be a special circumstance affecting the degree of care required also in the case of Brown and Brown v Gilbert-Scott and Payne.

The case involved a couple who wished to extend their house by building a conservatory. They employed an architect, a design was developed, and tenders were obtained. The tenders were too high. To save money a young, inexperienced builder with whom the architect had done some church work was, eventually, appointed.

The building work was carried out in a way that was far from satisfactory in terms of programme and quality. The architect's clients sued both the architect and the builder, submitting a long list of defects in design, inspection and workmanship.

The judge said:

> *It is not suggested that [the architect] knew [the contractor] to be inefficient or a bad builder. But, it is urged by the plaintiff that one of the circumstances against which this question of the appropriate degree or quality of his inspection is to be considered in this case was the youth and inexperience of [the builder]. I think that there is force in this submission: [the builder] was only young, [the architect] knew he had not been engaged in a job of this size before – although he had worked with him before and had found him to be a competent and reliable worker. But the work he had previously done for him before was on churches and work which was more properly described as that of stone-masonry rather than the more general work which was required in the construction of this conservatory. It cannot be said that [the architect] ought not to have recommended [the contractor] to do the work, but what can be said is that he was not a very experienced builder and that is a factor which [the architect] ought to have borne in mind when it came to the question of*

just how frequent or detailed his inspection of the work done by [the contractor] was to be. The [expert witness] called by the plaintiff ... said that given these circumstances, [the architect] should have watched [the builder] 'like a hawk'. I think that that is putting it 'too high' – but I do think that [the contractor's] age and inexperience was a factor which [the architect] ought to have borne in mind when he was discharging his obligation to inspect the work.

2.3 How often and for how long should the architect visit site?

In the Scottish case of *Jameson* v *Simon* (1899)[6] an architect had been employed in connection with the building of a house. The house was completed, but a month after the client moved in dry rot was discovered. Investigations showed that the rot originated from pieces of wood found in the sub-base below the ground floor slab. Two separate contractors had been involved in supplying and laying the sub-base. A mason was to have provided the bottom 2.5 ft – using waste stone arising from his work elsewhere on the house. A plasterer was then to have provided a 3 in deep upper layer of small stones, before laying the slab itself. It appears the quantity of waste stone generated by the mason turned out to be less than was needed to provide the bottom layer of sub-base. Although it is not clear who was directly responsible, the difference was made up by using general rubbish from the site – including pieces of wood. The architect had visited site on average once a week but had not been present during the period between when the laying of sub-base was started and when it was covered up by the slab. The client sued the architect.

The judge, the Lord Ordinary, Lord Kyllachy, said the architect:

undertook, like other architects, to grant certificates to the contractors upon which the contractors should be paid, and by these certificates he certified, if not expressly, by the plainest implication, that the work done had been done according to contract. Prima facie therefore, he must at least be held to warrant that in so far as he could ascertain by reasonable care and skill there had been no scamping of the work or serious deviation from the plans and specifications. That seems to be his prima facie undertaking, and I confess I am not prepared upon anything I have heard to put his responsibility lower. He was bound to supervise, and in doing so he was I think, bound to use reasonable care and skill, the burden being upon him to shew that with respect to any disconformity or default it was such as could not be discovered by reasonable care and skill. I cannot assent to the suggestion that an architect undertaking and being handsomely paid for supervision, the limit of his duty is to pay occasional visits at longer or shorter intervals to the work, and paying those visits to assume that all is right which he does not observe to be wrong.

The case went to the House of Lords, where it was accepted that in terms of the frequency and duration of his visits the architect had not deviated from the normal standards of the profession. However, the court agreed that it was not enough simply to visit the site at regular intervals and remain there for a given amount of time, but that the architect must make specific inspections of important parts of the works. Lord Justice Clerk said:

> ... as regards so substantial and important a matter as the bottoming of the cement floor of considerable area, such as this is shown by the plans to have been, I cannot hold that [the architect] is not chargeable with negligence if he fails before the bottoming is hid from view by the cement to make sure that unsuitable rubbish of a kind that will rot when covered up with wet cement has not been thrown in in quantities as bottoming contrary to the specifications. It is contended that the architect cannot be constantly at the work, and this is obviously true. But he or someone representing him should undoubtedly see to the principal parts of the work before they are hid from view, and if need be I think he should require a contractor to give notice before an operation is to be done which will prevent his so inspecting an important part of the work as to be able to give his certificates upon knowledge and not an assumption, as to how work hidden from view had been done.

The question has also been addressed in a number of more recent cases.

In an Australian case, *Florida Hotels Pty Ltd* v *Mayo* (1965),[7] a firm of architects was employed in connection with the design and construction of extensions to a hotel, including a swimming pool at the rear. No main contractor was employed. Instead, trade contractors and supervisory staff were employed directly by the hotel owner.

The hotel owner's managing director asked the architects to carry out their inspections regularly on Tuesdays and Fridays, when he would also be able to attend. The architects obliged.

The land on which the pool was constructed sloped away from the hotel. The structure of the pool-side furthest from the hotel was to comprise two concrete slabs arranged end to end. On a particular Friday morning one of the architects found that the formwork for the slabs was well advanced but not yet completed, and that no reinforcement was yet in place. He left site at lunchtime. During the remaining part of the day the formwork was completed and reinforcement fixed.

The reinforcement comprised a rectangular-patterned mesh. The engineer's specification required the mesh to be laid such that the bars at closer centres spanned continuously across the width of each slab.

The mesh was supplied in long rolls, with the bars at shorter centres arranged along the length of the rolls. This meant that to arrange the mesh as specified it would be necessary to cut the rolls into a number of short lengths and arrange them across the width of each slab. Instead, the mesh was simply cut into very long lengths, which were laid next to each other longitudinally from one end of each slab to the other such that no continuous bars spanned across the width of the slabs. The reinforcing strength of the mesh was thus reduced by 75%.

Without the knowledge of the architects, concrete was ordered on the Friday evening and delivered and poured on the Saturday morning.

On the following Tuesday the formwork was removed from one of the slabs, and removal of the formwork to the other slab started. The first slab collapsed, seriously injuring a workman who was removing the formwork from the second slab. The workman sued the hotel owner, who then joined in the architects, blaming them for not properly supervising the work.

The case went to appeal, where one of the judges, Chief Justice Barwick, said:

> ... there could not be any dispute that an architect with the obligation of supervising construction work is bound to supervise such an important step as the preparation of the formwork and the placement of reinforcement for the formation of such suspended concrete slabs as were intended in this case to form the eastern aprons of the swimming pool.
>
> But, because the supervision which the respondents were obliged to give was said ... to be 'periodical' [the architects] in effect claimed that as on all former occasions during the construction of the extensions to the hotel, presumably including also the earlier construction in connexion with the swimming pool, the foreman had alerted them to the time for the pouring of concrete, they were entitled to assume, in default of notification to the contrary, that concrete would not be poured between the days on which they made their routine inspections of the work ... I am clearly of the opinion that in law [the architects] were not so entitled. They were bound to supervise the work, inspecting it with due skill and care. There can be no doubt that due skill and care in this case required them to supervise the work done in preparation for the pouring of concrete to form these slabs. The facts of this case bring out starkly the importance of the performance of this obligation. In my opinion [the architects] were bound to take reasonable steps to ensure that they inspected the formwork and the placed reinforcement before concrete was poured and the work covered up. They do not satisfy this by relying on the workmen whose work they were employed to supervise: in particular, they were not entitled to assume from past satisfactory performances of the foreman, that they would be notified of the readiness of the work for inspection and of the time for the pouring of the

concrete. They were not engaged to supervise only such work as could be seen on the particular days of their routine inspections, or to attend to supervise only when advised that an occasion for supervision had arisen or was about to arise. They owed a duty to keep themselves informed of the progress of the work. They were bound, in my opinion, at least to have made reasonable arrangements of a reliable nature to be kept informed of the general progress of the work and, in particular, to be notified of the readiness of formwork and the placement of reinforcement for the pouring of concrete; these arrangements ought to have included clear and express instructions to the foreman that work of the kind in question must not be covered up till [the architects] had inspected it or, at the very least, had an adequate opportunity for its inspection.

The main points to note from the case are:

- the architect must inspect matters of importance
- the architect cannot rely on the contractor to advise him or her of when such inspections should be carried out.

In *Alexander Corfield* v *David Grant* (1992)[8] the defendant was a hotelier who, with his wife, ran a business that was highly recommended by a well-known hotel guide. They bought a listed private house into which, after carrying out some alterations, they wished to move their business. They appointed an architect, the plaintiff, and told him they needed the proposed building work completed ready to accept guests within eight months of the architect's appointment – in time to feature in the following year's guides. The vendors would not be vacating the building until two months after the architect's appointment. The defendant and his wife then expected to see building work start as soon as possible. Upon the architect's suggestion, in an attempt to make a prompt start on site, it was decided to negotiate with a builder, proposed by the architect, at the same time as the usual statutory consents were sought.

However, the unfolding of the job did not please the defendant, and he ended up refusing to pay fees owed to the architect. The architect sued, and the defendant counter-claimed, submitting a litany of alleged breaches of contract, among which was an accusation that the architect did not spend enough time on site, including that during the period of a particular month the architect spent less than an hour on site. Judge Bowsher QC, an Official Referee, said:

In the calendar month in question, the plaintiff personally went to the site on 5 occasions, he had 5 meetings in his office, and [the plaintiff's assistant] went to site twice. It is possible that the plaintiff's 5 site visits did not add up to an hour in total. The defendant is more likely to have been keeping a check on this than the plaintiff.

> What is adequate by the way of supervision and other work is not in the end to be tested by the number of hours worked on site or elsewhere, but by asking whether it was enough. At some stages of some jobs exclusive attention may be required to the job in question (either in the office or on site): at other stages of the same jobs, or during the most of the duration of other jobs, it will be quite sufficient to give attention to the job only from time to time. The proof of the pudding is in the eating.

In *Brown and Brown* v *Gilbert Scott and Payne* the question was considered again, although with an outcome not so favourable to the architect concerned:

> There was a certain amount of evidence and submissions were made as to the number of times an architect . . . in this contract should have visited the site. Here, [the architect] paid some 18 visits to the site. An attempt was made to show that each site visit would have taken some three hours in all, including travelling time from Oxford and that all in all he would have spent some 54 hours on site visits. I must say that I did not find this sort of analysis very helpful. It is not the number of visits made which is the way to judge the architect's performance of his duty to inspect. One must look to a whole range of factors such as the frequency of visits, the duration of each visit, what the architect did when he was there and how the visits fitted into the work which was being done by the builders.

> Again, it was suggested that inspection is not a very profitable part of the architect's engagement but I cannot see how that can affect his duty in respect of inspection. If an architect takes on the contractual obligation to inspect, then he ought to carry out this particular part of his engagement in a proper manner.

> It was further suggested that this site was quite a long way from [the architect's] office in Oxford and therefore that this must be taken into account. I cannot see why it should. As I have said, if an architect takes on a job which is some distance from his office, that should be no reason for offering any different sort of service – unless such was agreed between the parties.

The judge's general remarks are illustrated by his consideration of each of the specific defects of which the plaintiffs had complained. A good example is an apparent failure of the liquid-applied damp-proof membrane below the conservatory floor: fungus growth began to appear on the tiles a few months after they were laid. Upon investigation it had been found that the membrane had not been applied evenly to the correct thickness, and had not been lapped with the wall damp-proof course. In connection with the plaintiff's allegation that the architect, as well as the contractor, was to blame for the defect, the judge said:

> It is common ground between [the architect] and [the contractor] that the latter had telephoned [the architect] to tell him that he would be laying the [damp-proof membrane] either the next day or very shortly thereafter. However, [the architect] could not be there at the time when [the contractor] was going to do the work. He was conscious of the fact that [the plaintiffs] were getting impatient for work to be finished and he decided that it was not reasonable to hold up the tile laying. He therefore told [the contractor] to go ahead with the laying of the membrane. He said in his evidence that he took a calculated risk in not being there, but he said he trusted the contractor: it was not a difficult job and he believed that [the contractor] 'had the integrity' to do it properly.

The judge then reviewed the various arguments put forward by witnesses and concluded:

> In my judgement this was one of those critical stages in this job when the architect should either have been present or, at least, have satisfied himself that the work had been done properly. Directing myself in accordance with the various statements of principle I have set out above, I think that [the architect] failed in his duty to inspect during this critical phase of the works. As Judge Stabb QC said in Sutcliffe v Chippendale & Edmondson ... the architect's duty is to do all that is reasonably within his power to ensure that the work is properly done – he is engaged at this stage of the works (and for a considerable proportion of the overall fee) in order to protect the interests of his client. I conclude that it was not that urgent that [the architect] could not have said to [the contractor] that he was not able to come the following day to inspect the laying of the dpm and to have instructed him to wait for a few days until he could be present. Alternatively, if he was not able to be present whilst it was actually being laid, then he should have got there and inspected the work before the tiles had been laid. In view of the fact that [the contractor] thought he had to lay the tiles immediately after the laying of the screed, then he should have told [the contractor] to wait until the next week before it was done. [The architect] may well have been justified in trusting to the integrity of [the contractor], but there was also the issue of the experience and competence of this young builder to take into account and in this item, as with others, it was not so much a question of [the contractor's] integrity as his competence and experience.

> As to the apportionment of blame between the two Defendants for this item, I find that each is equally to blame for this defect and for the damages which flow from it ...

2.4 To what extent can the architect be expected to discover defects?

In the House of Lords case of *East Ham Borough Council v Bernard Sunley and Sons Ltd* (1965),[9] Lord Upjohn said:

> As is well known the architect is not permanently on the site but appears at intervals it may be of a week or a fortnight and he has, of course, to inspect the progress of the work. When he arrives on the site there may be many very important matters with which he has to deal: the work may be getting behind-hand through labour troubles; some of the suppliers of materials or the subcontractors may be lagging; there may be physical trouble on the site itself, such as, for example, finding an unexpected amount of underground water. All these are matters which may call for important decisions by the architect. He may in such circumstances think that he knows the builder sufficiently well and can trust him to carry out a good job; that it is more important that he should deal with urgent matters on site than that he should make a minute inspection on the site to see that the builder is complying with the specifications laid down by him ... It by no means follows that, in failing to discover a defect which a reasonable examination would have disclosed, in fact the architect was necessarily thereby in breach of his duty to the building owner so as to be liable in action for negligence. It may well be that the omission of the architect to find the defects was due to no more than an error of judgement, or was a deliberately calculated risk which in all the circumstances of the case was reasonable and proper.

Putting it simply: an architect will not be expected to spot every single minor defect.

The principles set forth by Lord Upjohn are exemplified by later cases such as *Victoria University of Manchester v Hugh Wilson* (1984).[10] An architect designed for the university a building of reinforced concrete, clad in brickwork and ceramic tiles. The tiles fell off, and the university sued the architect, the contractor, and the nominated tiling subcontractor.

The subcontractor went into liquidation before the trial, and the university and the architect settled on the fifth day of the trial. The trial continued as an action against the contractor, during which Judge John Newey QC, Official Referee, considered the standard of inspection required of an architect, saying:

> The subcontractors failed to butter the backs of the tiles correctly, so as not to leave voids, and they allowed sand to remain between some tiling courses. It would seem that on no occasion did the architects discover that the subcontractors were not doing their work properly. If the areas of tiling had been small and the subcontractors had worked on site for only a short period, I think that, bearing in mind the problems of supervision ... the architects might be excused for not having detected the subcontractors' failures.[11]

In another case, *George Fischer Holding Ltd v Multi Design Consultants Ltd and Davis Langdon & Everest* (1998),[12] under a design and build contract the plaintiff employed a contractor to provide a new warehouse building. The plaintiff also engaged an employer's representative, who was required by the contract of its engagement:

> To make visits to the site sufficient to monitor the contractor's workmanship and progress; to check on the use of materials, to check on the work's conformity to the specifications and drawings and to report generally on the progress and quality of the works having regard to the terms of the contract between the employer and the contractor . . .

Although the employer's representative was not an architect, and the contract was a design and build contract, the wording of the inspection duties to be performed is very similar to that used in architects' standard forms of appointment.

The contractor laid roofing panels with end laps that leaked, causing considerable damage. The plaintiff sued both the contractor and the employer's representative. Judge John Hicks QC, Official Referee, found:

> . . . [the employer's representative] made no visits to the roof whatever during the period when the panels were being laid and the lap joints formed, so they were undeniably in gross breach of duty. [The employer's representative's] only excuse for that omission was that access was not safe. That is obviously no answer; he was entitled to require the contractor to provide safe access.
>
> The only defence seriously advanced in respect of that breach was that the defective formation of the seals at the lap joints would not have been detected even had [the employer's representative] carried out inspection visits, because the work of making the seals would not necessarily or probably have been going on during the visits, and if it were the workman would have taken untypical care while under the eye of the employer's representative. That defence fails at every level. As to missing the relevant phase of the operation, first, it is clear on the evidence that on visits of the frequency and length which [the employer's representative] says he carried out elsewhere on site, and even without any special attention to this point, the likelihood is that he would on many occasions have had the opportunity of seeing lap joints formed and sealed. Secondly, the formation of the joints was so obviously crucial that even if the overall frequency of visits was not increased special attention should have been paid to ensuring that they fully covered this aspect. But, thirdly, since this whole discussion predicates the acceptance by [the employer's representative] of the very risky and inadvisable inclusion of lap joints in such shallow slopes, it was incumbent upon them to exercise the closest and the most rigorous inspection and supervision of the process. The last point also disposes of the suggestion that workmen will 'put on a show' – either they cannot do that all the time, or if they do that achieves the object anyway. Moreover it is in any event part of the necessary skill of a competent inspecting officer to detect and make allowances for such behaviour.

I therefore find both [the contractor and the employer's representative] responsible for the major contribution to leaking roofs made by the inclusion and defective construction of end laps in the roof panelling.

A final case, *London Hospital (Trustees)* v *T P Bennett* (1987),[13] is of interest because it addressed the question of the architect's liability when defects are deliberately concealed by the contractor.

Architects designed a 10-storey nurses' home of reinforced concrete, clad with brick panels. The panels were to be supported by reinforced concrete nibs projecting from the main structure. The contractor failed to set out the construction sufficiently accurately in the vertical direction, resulting in there being insufficient room to fit the required number of brick courses between nibs. The contractor then hacked away at the nibs until the brickwork would fit, in some cases reducing the nibs to no more than a few reinforcement bars protruding into thin air. Years after the building was completed a panel of brickwork began to bulge. Upon investigation it was found that the brickwork throughout the external elevations was so poorly supported as to be potentially dangerous. All of the brickwork was taken down, the nibs were repaired, and the brickwork rebuilt. The building owner sued the architect, the structural engineer and the contractor.

The plaintiffs accused the architect of failing to detect the destruction of the nibs. The judge found that during the architect's visits the contractor had, in a way that could not have been foreseen by the architect, deliberately concealed the damage to the nibs, and that therefore the architect could not be held liable for contributing to the damage. It was thereby established that there is no duty in law for an architect to prevent deliberate wrongdoing by a contractor – provided the wrongdoing cannot reasonably be foreseen.

2.5 What duties does the architect owe to the contractor in connection with defects?

The question can be answered by reference to two cases.

In *East Ham Borough Council* v *Bernard Sunley & Sons Ltd* a contractor was sued in connection with defective fixing of stone facing panels that started to fall from the exterior of a building after the issue of the final certificate. The contractor sought to avoid liability on the grounds that the architect had not spotted the defects and brought them to the contractor's attention during the course of construction. The case went to the House of Lords, where Lord Upjohn said:

It seems to me most unlikely that the parties to the contract contemplated that the builder should be excused for faulty work ... merely because the architect failed to

> carry out some examination which would have disclosed the defect. Even if the architect in failing to make the examination was in clear breach of his duty to his client, the building owner, I can see no reason why this should enable the builder to avoid liability for his defective work; the architect owes no duty to the builder except to issue certificates ... I cannot see why [the builder] should be allowed to escape from the ordinary consequences of his negligence when discovered years later, consequences which would undoubtedly flow if the building owner had not appointed an architect for his, the building owner's, protection.

In other words the architect is present for the benefit of the employer; he is not present to act as a scapegoat for the contractor.

Lord Pearson, as far as inspections made during the course of the works were concerned, agreed, saying:

> It seems to me unreasonable, too favourable to the contractors, to let them shelter behind the architect's failure to detect faults in the course of his visits during the progress of the work. The architect's duty is to the employers and not to the contractors, and the extent of his obligation to make inspections and tests depends on his contract with the employers and the arrangements made and the circumstances of the case. Prima facie the contractors should be and remain liable for their own breaches of contract, and should not have a general release from liability in respect of all breaches which the architect should have detected but failed to detect throughout the currency of the contract.

In a second case, *Bowmer & Kirkland Ltd* v *Wilson Bowden Properties Ltd* (1996),[14] a developer employed a contractor to build two office buildings. About six months after practical completion, leaking balconies and other defects became apparent. Legal action began with a writ issued on behalf of the contractor seeking orders that the developer place retention moneys in a separate trust. The developer counter-claimed damages arising from the defects. The contractor admitted that many defects existed, but said the cause of the defects was a combination of bad design and bad inspection by the architect (who was not joined in the action).

Judge Bowsher QC, Official Referee, said:

> ... it is alleged by [the contractors] that if there were any defects in workmanship on their part, those defects must have been apparent to the architect. Without spelling out the submission in explicit detail, [the contractors] appear to be saying that the architects had a duty to supervise their work and maintain quality control, and if the architects failed to maintain quality control, [the contractors] were to be excused from any defective performance of their duties under the contract. If that is their submission, it is wholly misconceived. The architects in this case were not under

a duty to supervise, and even if they had been, their duty to supervise would have been owed to the employers, not to the builders, and if there had been a breach of a duty to supervise, that would not have excused the builders from maintaining their own system of quality control ...

2.6 What duties does the architect owe to the contractor in connection with methods of working?

The question has been considered in a number of cases. In *Clayton v Woodman and Son (Builders) Ltd* (1962)[15] the contract works included the construction of a new lift motor room which was to abut a Victorian hospital clock tower. The new concrete floor of the motor room was to be housed into a chase to be cut into the wall of the clock tower. A bricklayer suggested to the architect that it would be difficult to form a waterproof junction between the new lift motor room roof and the existing clock tower, and that it would therefore be better to demolish the clock tower and form the motor room of entirely new work. The architect disagreed, and the work proceeded unchanged.

The builder cut the chase, but neglected to shore up the wall above. The wall fell and injured the bricklayer. The bricklayer sued the contractor, the employer and the architect – arguing that if the architect had allowed the bricklayer to demolish the tower as the bricklayer had suggested the bricklayer would not have been injured.

On appeal to the House of Lords it was found that the wall would not have fallen had the contractor taken proper precautions to ensure that the work would be carried out safely, and that the contractor alone was responsible for such matters. The judgment was summed up by Lord Pearson, who said:

The architect does not undertake (as I understand the position) to advise the builder as to what safety precautions should be taken or, in particular, as to how he should carry out his building operations. It is the function and the right of the builder to carry out his own building operations as he thinks fit, and, of course, in doing so, to comply with his obligations to the workman ...

... it cannot be right, in my view, to impose on the architect two conflicting duties in this situation: his duty to the owner to insist on the performance of the contract, and some other duty supposed to be owed to the builder or the builder's workman to make a variation to the specification in the circumstances of the case.

Secondly, it might be suggested that the fault of the architect was in not advising the builder, through his existing representative on site, the plaintiff, as to how the work required by the specification should be executed. If he had done so, the architect would have been stepping out of his own province and into the province of the builder. It is not right to require anyone to do that, and it is not in the interests of the builder's workpeople that there should be a confusion of functions as between the builder on the one hand and the architect on the other. I would hold that it was plainly not the architect's duty to do that. It will be observed that he had at any rate no pre-existing duty to do that. He was not asked to give any such advice and he did not profess to give any such advice, and I cannot see that it can be regarded as fault on his part that he did not step out of his province and advise the builder in what manner the builder should carry out his own building operations.

Thirdly, it might be suggested that the architect should have given a warning to the builder's workman ... as to how the work should be done or that there was some risk involved in doing it in a particular way. But there, also, it seems to me that that would have been stepping out of his own province and entering that of the builder. He was entitled to assume that the work would be carried out properly, that the builder knew his own business and would properly perform his own operations.

Fourthly, it might be suggested that the architect should have stopped the progress of the work. There again there was no need to do that unless he could assume that the builder did not know his own business and was not going to do his own work in the right way.

Lord Pearson finished by implying that had the work been inherently impossible to carry out safely, or had the architect expressly agreed to take responsibility for safety precautions, it is likely he would have been found negligent.

In *Clayton* v *Woodman* the architect quite correctly refused the contractor's request to vary the works. In contrast, the architect in *Clay* v *A J Crump & Sons Ltd* (1964)[16] was rash in his willingness to vary the works in compliance with a request from his client.

With a view to extending his premises an owner acquired property adjoining the site of his garage business. An architect was employed. On the newly acquired part of the site it was proposed to demolish a number of existing buildings, carry out excavations to reduce levels, and then erect new buildings. To save time the demolition and excavation work was let under a separate contract to be carried out before the main building works. Demolition and main building contractors were appointed. The architect's demolition and excavation contract drawings required the demolition contractor to leave in place 5 ft wide retaining banks of earth at the base of various walls that were to be retained, and which would otherwise be undermined.

While work was under way the garage owner became concerned that the demolition of a particular wall, which had just been started, would allow anyone who wandered onto the site to gain unauthorised access to his existing premises. The owner approached the demolition contractor's foreman and told him to stop demolishing the wall. The owner then telephoned the architect, who undertook to deal with the matter formally.

The architect asked the demolition contractor by telephone whether it was safe to leave the wall standing. The demolition contractor said he thought it was. The architect instructed that the wall be left in place, but did not inspect the condition of the wall at the time, make any specific enquiries as to how far excavation had proceeded, or instruct that a 5 ft retaining strip be left at the base of the wall to match that at the base of other walls to be kept.

Later the architect went to site but did not bother to check that the wall was safe: the wall had been left standing on a precipice about 2 m high and was not bonded to abutting walls.

Six weeks after demolition of the wall had been stopped, some time after the demolition contractor had left site, the building contractor erected within 2 ft of the base of the wall a hut to be used for the storage of tools, and as a place for the people working on site to eat meals. The wall collapsed on the hut, killing two men and injuring another. The injured man sued the architect, the demolition contractor and the building contractor. The court found all three liable. Blame was apportioned as 42% to the architect, and lesser amounts each to the demolition and building contractors.

The case involved poor design and poor inspection, and is a tragic illustration of the potential dangers inherent in making, at the request of others, heat-of-the-moment variations to a design during construction.

Judge Stabb QC in a third case, *Oldschool* v *Gleeson Construction Ltd* (1976),[17] considered comprehensively the question of the consultant's duties with respect to methods of work. Although the case concerned an engineer, the principles of the judgment apply equally to architects.

A party wall collapsed during demolition and excavation works. The employer sued first the contractor and second the engineer. The contractor instituted third party proceedings against the engineer, claiming that the wall had collapsed because the engineer's design was impossible to build without the wall collapsing, and that the engineer's supervision of the works was inadequate, such that the wall would have collapsed even if the design had not been impossible to build.

The judge considered the views of the contractor's and engineer's expert witnesses as follows:

> [The engineer's expert was] insistent that the manner of the execution of the works is a matter for the contractors. He considered that the consulting engineer is in no position, for instance, to require the contractors to comply with any particular sequence of works; he has no right, let alone duty, to involve himself in the work of the contractors. Of course he would interest himself in their work, would offer advice to assist the job to go better and would certainly not turn his back on a situation that he could see was likely to give rise to danger to life. Equally he would intervene if he could see imminent damage to property. Those are matters of common sense; but that is a very different matter from assuming responsibility for the method of work to be adopted by the contractors.
>
> In my judgement [the engineer's expert's] view is the right one. I do not think that the consulting engineer has any duty to tell the contractors how to do their work. He can and no doubt will offer advice to contractors as to various aspects of the work, but the ultimate responsibility for achieving the consulting engineer's design remains with the contractors ... If the contractors had said, for example, that they planned to excavate first down to footing level along the whole length of the party wall and thereafter to excavate the rest of No 31, the consulting engineer might well have pointed out the undesirability or even the danger of adopting that course; but I do not think that he was under a duty to direct the contractors, for instance, to excavate in strips up to the party wall. It was the responsibility of the contractors to decide upon the method and sequence of excavation so as to achieve the consulting engineer's design; but if, for example, they planned to excavate the hoist pit without any temporary support, and so informed the consulting engineer, then as a matter of common sense the consulting engineer would intervene to prevent that which was described as 'an act of incredible folly'.
>
> From the evidence which I have heard and from the contemporaneous documents I am satisfied that the second defendants adequately fulfilled their duty of supervision. [The engineer] persistently drew [the site agent's] attention to the inadequacy of the shoring, although he was not, in my view, duty bound to do so. He warned him of the risk that he was running. He emphasised the necessity to blind the excavated ground by the party wall at the earliest opportunity. He told [the site agent] ... before any excavation of the hoist pit had started, to put in the sheeting or strutting, in the form of precast concrete planks, which were on the site, and not to excavate further until this was done. In the circumstances he had no reason to foresee that further excavation would be carried out until this was done. He personally visited the site seven times in the six weeks period [before the date of collapse] and in spite of [the site agent's] evidence to the contrary, I believe that [the engineer] advised and warned [the site agent] in the manner which he described ... [The contracts

manager] and indeed [the contractor's expert] agreed that, if the contractor had any doubts about how the excavation should be done or how the temporary support should be set or how the underpinning should be carried out, then they should have asked the consulting engineer, but this they never seem to have done.

... What is said [by the contractor's expert], however, is that when the consulting engineer knows or ought to know that the contractors are heading into danger whereby damage to property is likely to result, then he owes the contractors a duty of care to prevent such damage occurring. If he sees the contractors not taking special precautions without which a risk of damage to property is likely to arise, then he the consulting engineer cannot sit back and do nothing. I am not sure that the consulting engineer's duty extends quite that far but, even if it does, I do not believe that he is under a duty to do more than warn the contractors to take the precautions necessary ...

The judge continued:

It seems abundantly plain that the duty of care of an architect or of a consulting engineer in no way extends into the area of how the work is carried out. Not only has he no duty to instruct the builder how to do the work or what safety precautions to take but he has no right to do so, nor is he under any duty to the builder to detect faults during the progress of the work. The architect, in that respect, may be in breach of his duty to his client, the building owner, but this does not excuse the builder for faulty work.

I take the view that the duty of care which an architect or a consulting engineer owes to a third party is limited by the assumption that the contractor who executes the works acts at all times as a competent contractor. The contractor cannot seek to pass the blame for incompetent work onto the consulting engineer on the grounds that he failed to intervene to prevent it.

2.7 To what extent is the architect liable for the performance of the clerk of works?

An early case that addressed the relative responsibilities of consultant and clerk of works is *Saunders and Collard* v *Broadstairs Local Board* (1890).[18] The case involved engineers who were employed by a local authority in connection with the design and construction of a drainage scheme. The local authority were to appoint a clerk of works. The engineers apparently raised doubts about the competence of the clerk of works put forward by the local authority – but he was nevertheless employed. Upon completion the local authority alleged, among other defects, that a section of drain had been laid to incorrect levels such that it permanently contained sewage and water. The local authority blamed the engineers for not checking levels during construction. The engineers said that it was the job of the clerk of works to check the levels, that they had explained to him how to do it, and that if any levels were wrong it was due only to his incompetence.

The judge, Mr Ridley, an Official Referee, took the view that the engineers were negligent in respect of the matter on two counts:

- they relied on the clerk of works to check levels when they suspected he was incompetent
- they should in any case have themselves checked something as important as the levels.

In another early case, *Lee v Bateman* (1893),[19] a firm of surveyors were employed in connection with the renovation and restoration of a kitchen wing following a fire in a mansion owned by Lord Bateman. A clerk of works was appointed by Lord Bateman.

Upon completion it was found that some beams were rotten, and it was alleged that the rot should have been spotted by the surveyors in time for replacement to be carried out during the course of the main building works. While work was in progress the surveyors had asked the clerk of works to check whether or not the beams needed to be renewed. The clerk of works had advised that replacement was not necessary – but the surveyors had not themselves checked. The judge, Mr Justice Cave, directed the jury that such an important question was a matter for the surveyors and should not have been delegated to the clerk of works.

A further case of interest is *Leicester Guardians v Trollope* (1911).[20] A firm was appointed to carry out architectural services in connection with the design and construction of a large addition to an infirmary. A clerk of works was appointed by the architect's client.

Instead of constructing the ground floor as specified, the contractor drove wooden stakes into the ground, suspended timber joists from the stakes, laid a poor-quality felt below the joists, and poured concrete to the level of the tops of the joists, without compacting it. Dry rot developed in the stakes at ground level, and spread from them via the joists to other parts of the building. It was discovered a few years after the building was occupied.

The client sued the architect. The architect blamed the clerk of works. The judge, Justice Channell, said:

It was clearly the duty of the clerk of the works to attend to the laying of concrete in accordance with the design, but does that relieve the [architect]? To my mind there is little difficulty in deciding the point. The position of the architect and of the clerk of the works was made quite clear. The architect could not be at the works all the time, and it was for that reason that the clerk of the works was employed to protect the building owner. But what is a matter of detail? The laying of this concrete was, in my view, a very important matter in relation to the building. It requires no expert to tell one that when a floor is put down on earth there must be protection against damp. Here the plan devised was not uncommon, and was an essential part of the design. The architect admits that he took no steps to find out whether it was carried out or whether it was not. A large area had to be covered. In some parts the concrete as laid was fairly good, but over the greater part it was all rotten. If the architect had taken steps to see that the first block was all right, and had then told the clerk of the works that the work in the others was to be carried out in the same way, I would have been inclined to hold that the architect had done his duty; but in fact he did nothing to see that the design was complied with. In my view this was not a matter of detail which could be left to the clerk of works.

Much more recently, in *Kensington A H A v Wettern Composites*, the judge said:

... the appointment of a clerk of works, whilst a factor to be taken into account, does not reduce the architect's liability to use reasonable skill and care to ensure conformity with design, as opposed to mere detail. The distinction may not always be easy.

The judge went on to quote Justice Channell in *Leicester Guardians* v *Trollope*, and concluded that the architects in *Kensington* v *Wettern* failed to carry out their inspection duties as they should have done.

The judge ended his judgment with a consideration of the possibility of contributory negligence on the part of the structural engineer and the clerk of works. The judge said:

The position of the clerk of works calls for different considerations. He is not ordinarily professionally qualified, but has practical knowledge of the building trade. He has been aptly described as the Regimental Sergeant-Major. It is accepted that he acts as the eyes and the ears of the architects, and has a responsibility to keep the architects informed as to what is or is not happening on site. He is also described as employed 'to act solely as inspector on behalf of the employer under the direction of the architect'.

The judge, finding the clerk of works had indeed been negligent, homed in on the question of vicarious liability:

Counsel were not able to refer to any decision in the common law world where consideration has been given to vicarious liability for the negligence of a clerk of works employed by a building owner yet under the architect's direction and control ...

If the plaintiffs had intended to abrogate the relationship of master and servant as between themselves and the clerk of works then one would have expected the terms of appointment to have made that plain ... There is no such evidence in this case. In my view the plaintiffs are vicariously liable for the negligence of the clerk of works.

I have reached the conclusion that the clerk of works' negligence whilst more than minimal is very much less than that of the architects. If I may adapt the military terminology: it was the negligence of the Chief Petty Officer as compared with that of the Captain of the ship. I assess the responsibility as to clerk of works 20%, as to the architects 80%.

In simple terms:

- The employment of a clerk of works can reduce the architect's liability in relation to inspecting matters of detail, but cannot reduce the architect's liability in relation to 'important matters'.
- The architect will not be held liable for the negligence of the clerk of works unless the clerk of works is employed by the architect, or it can somehow otherwise be proved that all control of the clerk of works has been divested to the architect or that the clerk of works is the architect's servant in law.

Notes

1 Sutcliffe v Chippendale and Edmundson (1971) 18 BLR 149.
2 Brown and Brown v Gilbert Scott and Payne (1992) 35 Con LR 120.
3 Consarc Design Ltd v Hutch Investments Ltd (1999) 84 Con LR 36.
4 The Kensington Chelsea and Westminster Area Health Authority v Wettern Composites and Others (1984) 31 BLR 57.
5 Defects discovered included:
   - inadequate vertical support to 85% of the mullions due to damaged concrete corbels, loose metal corbels or ill-fitting panels
   - inadequate horizontal support to 25% of mullions due to no cramps being fitted, cramps being bolted to the structural frame rather than fitted into abbey slots, the absence of dowels, or inadequate length or positioning of dowels such that dowels were not engaged in mullion holes
   - mortar left in horizontal joints preventing relative movement between mullions.
6 Jameson v Simon (1899) IF Court of Session 1211.
7 Florida Hotels Pty Ltd v Mayo (1965) 113 CLR 588.
8 Alexander Corfield v David Grant (1992) 59 BLR 102.
9 East Ham Borough Council v Bernard Sunley and Sons Ltd (1965) 3 All ER 619.

10 Victoria University of Manchester v Hugh Wilson (1984) 2 Con LR 43.
11 The judge went on to say that because the areas of tiling were extensive and were carried out over a long period of time the architects were negligent.
12 George Fischer Holding Ltd v Multi Design Consultants Ltd and Davis Langdon & Everest (1998) 61 Con LR 85.
13 London Hospital (Trustees) v T P Bennett (1987) 43 BLR 63.
14 Bowmer & Kirkland Ltd v Wilson Bowden Properties Ltd (1996) Unreported 11 January 1996.
15 Clayton v Woodman and Son (Builders) Ltd (1962) 2 All ER 33.
16 Clay v A J Crump & Sons Ltd (1964) 3 All ER 687.
17 Oldschool v Gleeson Construction Ltd (1976) 4 BLR 103.
18 Saunders v Broadstairs Local Board (1890) Hudson's BC 4th edn Vol 2 164.
19 Lee v Bateman (1893) *The Times* 31 October.
20 Leicester Guardians v Trollope (1911) 75 JP 197.

3 Practice management matters

3.1 How can the architect contribute to the achievement of high-quality work on building sites?

The biggest contribution the architect can make to the achievement of high-quality building work is not by inspecting work in progress but by ensuring that the builder is provided on time with clear, complete and properly coordinated production information. If information is poor, unfinished, provided at the last minute, or late the contractor will:

- spend time chasing the architect for information or clarification, instead of carrying out his own inspections of work in progress
- be unable to foresee and resolve technical problems before work starts
- be unable to organise subcontractors and materials most efficiently
- be tempted to guess or make assumptions about construction details
- become frustrated and demotivated.

Any of the above will inevitably lead to a lowering of construction quality, to poor cost control, and to potential delays. Furthermore, it is likely that the architect will be so busy answering contractor's queries or sorting out matters of detail design that there will be inadequate time for carrying out the inspection duties for which he or she may have been appointed. The quality of production information therefore has a greater influence on building quality than does the inspection of work in progress.

Nevertheless, even given adequate production information, there is no doubt that:

- the complexity of building operations
- the potential for misinterpretation of even the best production information, and
- human fallibility

will mean that in all but the most exceptional cases inspection services properly carried out by the architect, alone or with others under the direction of the architect, will significantly increase the quality of the completed building.

3.2 The architect's appointment

Under past standard forms of appointment the architect was not required to inspect every part of a building as it was being built, and would not normally have been paid fees to enable him or her to do so. The architect should therefore have:

- checked essential parts of the design as they were carried out
- carried out periodic and spot checks of the construction generally
- during the course of the above compared progress on site with contract completion dates and construction programmes, and checked that progress was regular
- checked that the contractor was himself maintaining adequate arrangements for supervision and quality control.

Unfortunately, at the time of writing, when the services to be performed under an architect's appointment need to be defined with an unprecedented degree of precision, the standard forms of appointment most recently published by the RIBA are, in relation to inspection duties, exceptionally vague and unhelpful.

In the past, judges have occasionally referred to the *Architect's Job Book* for insight into the services that could be expected of an architect as standard. However, although the most recent addition of the *Architect's Job Book* describes a number of site inspection activities, it does not say which, if any, the architect should normally carry out. Instead, the architect is advised to 'visit site as provided for in the Agreement with the client'. The advice effectively makes optional the various site inspection activities that are described. Consequently, it is no longer possible to refer to the *Architect's Job Book* as a benchmark.

It seems to have been concluded that the variety of procurement arrangements now in current use has made it impossible to define standard inspection services, and that it is therefore necessary to define separately for each project the particular inspection services to be performed. Presumably, such definitions should somehow be attached to standard appointment documents, or be set out in covering letters.

Such a situation is potentially dangerous for the architect because without standard definitions of services, prepared with the assistance of legal experts, the architect is vulnerable to accepting more liability than was bargained for. Architects may, for example, think that they have agreed to attend site infrequently, or occasionally, just to answer contractor's queries, and may accordingly accept fees commensurate with a minimal amount of time being spent on site. However, they may later find that because of unfortunate wording – or lack of wording – in their appointment document, they have almost the same liability as if they had undertaken to provide full inspection services.

Architects must therefore ensure that within their appointment documents the inspection services to be performed are defined with the utmost care, taking all potential liabilities into account – including liabilities to third parties such as funders, and future owners or tenants. Current and previously published standard forms of appointment, and the *Architect's Job Book,* will undoubtedly offer some guidance, but their use cannot guarantee the avoidance of all pitfalls. Legal advice in any but the simplest of situations should therefore be considered.

In practice, before finalising his or her appointment, the architect should discuss with the client the options for inspection services, advise the client of the advantages and disadvantages of each, and then agree with the client the services to be provided. The architect should record in writing all that was said during discussions, and incorporate into the formal appointment documents as complete and clear a description of the services to be provided as possible – taking legal advice as necessary.

If the architect has been appointed for services that do not include site inspection, the architect's appointment documents should make it clear that the architect will not attend site, even in response to queries from the contractor. In such a case the architect should of course ensure that he or she does not attend site to deal in any way with the contractor, and is best advised to avoid the site completely until all building work is finished.

3.3 Should the architect offer to carry out reduced or partial site inspection services?

It may be tempting for a client to try and reduce expenditure on fees by asking his or her architect not to visit site as often as the architect advises. The architect should be wary of agreeing to such an arrangement, as:

- it is inconsistent with a duty to certify payments: if an architect were to certify payment for work that had not been properly inspected it would probably be professional negligence – whatever was agreed with the client. In any case, certificates based on inadequate inspections are of little value to a client.
- even if an architect had no responsibility to certify payments it is likely that reducing the number of visits would give the architect much less time on site in which to spot errors in construction – for which it is likely the architect would still be partially blamed. In defence against such accusations it would be difficult to distinguish between which defects, in the light of the reduced services, the architect could be excused from noticing, and which he or she would still be obliged to discover.

Such arrangements should be avoided. If they are not avoided, difficulties will be experienced in reducing the architect's liabilities.

3.4 Time allowance

The amount of time that is required or which should be allowed for properly inspecting work in progress depends very much upon the nature of the job in question. It is certain, however, that if the architect is sued for negligent inspection it would be no defence to argue simply that the fee was not enough to cover the costs of carrying out inspection duties adequately.

It is therefore essential that, before agreeing fees with a client, the architect gives careful consideration to the time required to inspect the works. Factors that should be taken into account include:

- the expected contract period
- the size and complexity of the job
- the resources available in the architect's office
- whether or not site staff are to be employed, by whom, and the amount of time that site staff will be spending on site
- whether or not the contractor and the architect have worked together before
- the capabilities of the contractor, as demonstrated both before and during the contract period
- the extent of innovation in the design
- the standard of quality required by the contract documents
- other relevant terms of the contract, such as the frequency with which certificates for payment are to be issued
- the health and safety implications of the design
- the time required to carry out other construction stage activities, such as: attending and writing minutes of meetings; answering queries; preparing and issuing additional production information, instructions and certificates; coordinating the work of other consultants; reporting to the client; and general administration.

A record of all allowances made for site visits should be kept on file, both for the purpose of monitoring time spent, and for use as justification of additional fees for necessary site visits that could not reasonably have been foreseen at the time of the architect's appointment.

The situation should be monitored throughout the design stages of the job. Should the amount of time required for site inspections increase as a result of client variations to the design, or other circumstances additional fees should be negotiated. In particular, great care should be taken not to use up too much time in the design and production information stages of the job, leaving insufficient fees to carry out construction stage duties properly. (Times have changed dramatically since Lord Kyllachy in *Jameson* v *Simon* referred to the architect 'undertaking and being handsomely paid for supervision'.) Once on site, as with every other stage of a job, the architect should continue to monitor costs and take action as necessary to ensure that his or her work is carried out as efficiently as possible.

3.5 Who should do the inspecting?

Properly inspecting work in progress requires knowledge and experience, and is not a job to be delegated to junior staff with insufficient of either – whether to save money or because senior staff are too busy. However, junior staff must be given plenty of opportunities to attend site inspections in order to learn the necessary skills for the future. On a bigger or more complex project job architects should before and after visiting site report to the project principal, who should discuss with the job architect matters of importance. All of the above should be taken into account when estimating the time required, and the fees to be charged.

3.6 Should there be a practice site inspector?

In general, it is better not to devolve all site inspection work to a practice site inspector.

It may be argued that within a practice all site inspections should be carried out by a single architect – or, within a larger office, by a team of architects – and that benefits would be derived from the specialist experience that would be gained by such an architect or team. However, in the overwhelming majority of circumstances it is likely that such divorcing of inspection from design, drawing and specification processes would result in a significant risk that important design intentions – both technical and aesthetic – would be overlooked, or that their importance would be undervalued, leading to disappointing or even defective work on site.

Furthermore, from the point of view of the architect doing the design and production information work, the essential experience of seeing the implications of his or her drawings at first hand on site would, under such a system, be lost. In any case, site experience gained by each member of a practice can and should be shared with the practice as a whole. It can be done, for example, by holding occasional lunchtime discussions, or by compiling a book of points to look out for.

Circumstances under which a specialist inspector or team of inspectors may be of benefit are when a relatively large number of similar, system-built buildings or parts of buildings are to be constructed in succession – such as fitted kitchens, conservatories, temporary accommodation, prefabricated pods, or certain types of industrial building.

3.7 Other consultants' appointments

Before entering into an agreement to provide site inspection services the architect should advise the client on the need to appoint other consultants to inspect elements of the works that are beyond the scope of the architect's knowledge. All but the simplest of structural and building services work is likely to require inspection by consultant engineers.

3.8 The tender documents

The tender documents play a critical role in the achievement of high-quality work on building sites. They should of course, in objective terms, make clear the quality and standards of work required. However, during the preparation of tender documents proper consideration should be given to a number of matters that can indirectly influence both the quality of building work, and the effectiveness and efficiency with which the architect can carry out inspection duties. In particular:

- Checking that the contractor is attending properly to his own quality control responsibilities can make an important contribution to the quality of the finished product. Such checking can be made easier if the contractor is required by the tender documents to provide:
  ○ with his tender – or at the latest before starting on site – a statement describing the organisation and resources by which the contractor proposes to control the quality of the work
  ○ during the course of the work records of inspections, tests, and actions taken following the discovery of defective work.

  On larger jobs the contract documents should require the contractor to submit to the architect routine reports on the contractor's quality control activities, and on the progress of the works. The reports could be submitted as part of the contractor's report at each site progress meeting. The contractor should also be required to provide documentary evidence of the quality of materials and goods being incorporated into the work.
- The contractor should be expressly required by the tender documents to provide:
  ○ a competent person-in-charge
  ○ competent trade supervisors
  ○ operatives appropriately skilled and experienced for the type of work they are carrying out

- proper protection of stored materials and finished work (which should be explicitly described in relation both to vulnerable or valuable individual items, and to protection generally)
- welfare facilities of high quality, which are to be kept properly working and clean
- adequate measures to maintain the security of the site (with any special requirements made clear)
- a site kept tidy and clear of all debris.

• The monitoring of progress (and the assessment of claims associated with delays) can only be carried out properly if the contractor provides:
- a programme that shows periods allowed for planning and mobilisation; subcontract works; ordering of long lead-in materials and components; preparation of production information to be provided by the contractor or subcontractors, including allowances for checking and commenting by consultants, revising, and reissuing for acceptance; testing and commissioning; and work in connection with provisional sums
- a record of progress, displayed on a copy of the programme kept on site
- weekly records of the number and description of all persons employed on or in connection with the works
- weekly records of the number, type and capacity of all mechanical and power operated plant employed on the works
- a weekly record of materials delivered to site
- an accurate record of daily maximum and minimum air temperatures, and the number of hours per day in which work is prevented by adverse weather.

A requirement to provide the above should be explicit within the tender documents. Consideration should be given to including a proviso that the employer will not enter into contractual arrangements until a satisfactory contractor's programme is provided.

• One of the best ways of controlling the quality of materials and assembled elements in buildings is by the use of samples and mock-ups. They should be used to:
- check materials supplied by the contractor by comparison with samples of known quality
- finalise matters of appearance by comparing on site sample panels showing, for example, the effects of different brickwork mortar colours or external render textures
- provide a control sample of a building element – such as brickwork, a flooring finish, or visual concrete – against which work in progress can be checked
- allow technical and visual problems to be brought to light and resolved by the construction of a mock-up – of a cladding or complex masonry detail for example – before carrying out work on the building itself.

The provision of such samples and mock-ups costs money and takes time, so it is important to ensure that adequate provision is made for them in tender documents. If a building has repetitive elements it may be possible to use the first element for control purposes. Control samples and mock-ups should be kept on site for reference until the relevant building work is finished.
- The power to instruct opening up and tests as necessary to carry out spot checks of work covered up can significantly contribute to the architect's ability to control quality. The provision of such power can in itself serve to increase the contractor's vigilance. The exercise of such power does, however, cost time and money, so it is essential that the tender documents require the contractor to make appropriate allowances.
- The inspecting architect's life can be made easier if the tender documents require the contractor to give reasonable notice to the architect before:
  ○ scaffolding or other access is removed
  ○ important tests and inspections are carried out (although the architect should not rely on the contractor to give such notice).
- Financial control will be improved if the contractor is required by the tender documents to provide reasonable notice to the architect and quantity surveyor of the start of any work for which daywork sheets are to be submitted, and to provide an accurate record of time spent and materials and plant used.

3.9 Clauses 2.1 and 8.2.2 of JCT 98

Clause 2.1 of JCT 98 is a fundamental clause. Most building contracts contain similar clauses. It states:

> The Contractor shall upon and subject to the Conditions carry out and complete the Works in compliance with the Contract Documents, using materials and workmanship of the quality and standards therein specified, provided that where and to the extent that approval of the quality of materials or of the standards of workmanship is a matter for the opinion of the Architect such quality and standards shall be to the reasonable satisfaction of the Architect.

The clause establishes the basic obligation of the contractor to carry out the work in accordance with the contract documents. It also appears to allow the required quality and standards of some materials and workmanship to be specifically described, and the required quality and standards of other materials and workmanship to be left simply to the opinion of the architect. However, except that insofar as the quality and standards of materials and workmanship are specifically described in the contract documents the architect has no power to insist on anything better, the courts have found that the quality of all materials and the standards of all workmanship are matters for the opinion of the architect – whether or not any specific descriptions of materials and workmanship are contained in the contract documents.[1]

The courts' interpretation of clause 2.1 of JCT 98 may seem for most intents and purposes to be merely academic, but the interpretation affects insidiously the meaning of clause 8.2.2, which states:

> *In respect of any materials, goods or workmanship, as comprised in executed work, which are to be to the satisfaction of the Architect in accordance with clause 2.1, the Architect shall express any dissatisfaction within a reasonable time from the execution of the work.*

In other words, given that the courts have established that all work is subject to the opinion of the architect, if the architect fails to spot a defect 'within a reasonable time' the contractor may escape liability. Such an outcome would of course be extremely unjust. Although the question has not yet been tested in the courts, the best policy when JCT 98 is to be used is to have stated in the tender documents that clause 8.2.2 will simply be deleted from the contract. Fortunately clause 8.2.2 is unique to JCT 98; similar clauses do not appear in other contracts.

3.10 Is a clerk of works or site architect required?

The architect is usually required to make only periodic inspections, rather than frequent or constant inspections, of work in progress. However, the size and complexity of a project, the speed of construction, the type of contract, or other factors may demand more inspection work than can be carried out by an architect visiting site periodically. Alternatively, the distance between the architect's office and the site may make journeys between the two at sufficiently frequent intervals impractical. Under such circumstances the standard forms of appointment require the architect to advise the client on the need to appoint site inspectors.

In simple terms, the job of site inspectors is to look, and report to the architect what they see. Under conventional building contracts site inspectors have no power to issue instructions to the contractor, to approve any of the work, to issue certificates, or to carry out any of the other administrative duties that fall to the architect.

The appointment of site inspectors allows more of the work to be checked, and allows it to be checked in greater detail, but does not relieve the architect of his or her own periodic inspection duties. Furthermore, if a clerk of works or other inspector is employed by the architect rather than by the client the architect will be liable for the inspector's negligence, and may unwittingly have made him or herself liable for duties beyond mere periodic inspection. The architect should therefore resist either employing, appointing or paying site inspectors such as clerks of works, or appointing a member of his or her own staff as a full- or part-time site architect. All such matters are best left entirely to the client.

Site inspectors may comprise one or more of the following:

- a 'traditional' clerk or clerks of works
- a specialist clerk of works
- a site inspector, who is not a clerk of works but who will carry out a job similar to either of the above
- a site architect
- a site services or structural engineer.

As with most matters the appropriate arrangement will depend upon the nature of the project.

The size of a job may itself demand more inspection than can be provided by the architect carrying out periodic visits. If such a job involves largely traditional building trades carried out under a conventional form of contract it is likely, up to a certain size, that the appointment of a traditional clerk of works alone will suffice. If the job involves state-of-the-art construction methods for which a clerk of works with the necessary specialist expertise cannot be found, then the services of a site architect or other inspector will be required. A clerk of works may also be appointed.

In other cases the speed of construction may make necessary the permanent presence of a site inspector. The presence of specific site staff may also be explicitly required by particular types of contract or warranty.

If the site is too far from the architect's office for the architect to visit as often as is necessary to inspect the works properly – but the job is not big enough to justify the employment of a full-time site inspector – a local part-time clerk of works or other site inspector will need to be appointed. Suitable arrangements may include the appointment of a local architect or surveyor to carry out part-time or periodic inspection duties.

Of increasing importance is the need to ensure that mechanical and electrical installations are inspected by those suitably qualified to do so. Even small residential refurbishments now commonly involve the installation of:

- boosted water supplies
- unvented domestic hot water systems

- sealed central heating systems
- pumped showers
- fire and security alarm and detection systems
- built-in hi-fi systems
- lighting control systems
- integrated telephone and door entry phone systems
- multiple telephone line services providing for TV, computer, alarm, and telephone systems.

Other work is likely to include even more complex systems. If an architect is certifying payments for work that includes the installation of such systems he or she must ensure that adequate inspections are carried out. It is likely that in most cases the architect will have no more specialist knowledge of such systems than the client. The architect's professional indemnity insurance may also not cover the inspection of such work. If, under such circumstances, the appointments of other consultants on the design team do not allow for adequate inspection of such specialist work the architect should advise the client to employ suitably qualified staff to carry out the necessary inspections.

Similar considerations should also be given to complex structural work.

In view of the above the architect should always give thought to whether or not site staff are needed, and ensure that advice is given to the client as required by the architect's conditions of engagement. Such advice should be given early enough to ensure that plenty of time is allowed for suitable people to be found and appointed. By failing to advise the client when site staff should be employed, the architect may inadvertently be taking responsibility for additional or specialist inspection duties. If, in spite of the architect's advice, the client decides against the appointment of necessary site inspectors the architect should ensure that the client has been warned of the associated risks – and that such warnings are recorded. Even then it is likely that the architect will have to carry out more frequent inspections him or herself.

The architect should also ensure that the building contract allows for inspections by the site inspectors who are to be appointed.

3.11 Finding a clerk of works

There are a number of ways to find a clerk of works or site inspector, including:

- contact from a previous job
- introduction by the client or another consultant on the design team

- advice from local consultants, clients, or contractors
- directories published by the Institute of Clerks of Works and other professional institutes
- advertising in local or trade press
- a combination of the above.

Before a prospective clerk of works or site inspector is employed, both the architect and the client must be satisfied that the clerk of works or site inspector is suitably qualified and able to do the job. Personality is as important as the right kind of experience. A clerk of works who is likely repeatedly to irritate the contractor or who is unsympathetic to the priorities of the project should not be employed. Unless a suitable and willing candidate is already known, it will be necessary to interview several. The architect should be present at the interviews.

The clerk of works and other site inspectors should be appointed as soon as possible before starting on site to enable them to:

- assist in the finalising of production information by providing insights derived from specialist or local knowledge
- familiarise themselves fully with the contract documents
- get to know the personalities and procedures associated with the project.

The appointment of the clerk of works or site inspector should be thought through and carried out with great care. In doing so the architect should refer to the latest editions of the *Architect's Job Book* and the *Clerk of Works Manual,* both published by RIBA Publications. The client should confirm working hours, holidays, payment for time spent travelling and for expenses, and other terms of employment.

Notes

1 Crown Estate Commissioners v John Mowlem (1994) 70 BLR 1.

4 As work is about to start

4.1 Importance of pre-planning and prioritising

It is vitally important that, before work starts, proper consideration be given to inspection.

The safest approach is to study the production information and prepare a list of all the parts of the design that it is essential be inspected. A plan should then be set up showing when on the basis of the contractor's programme it is expected the listed parts will be built, when they should be inspected, and when checks of the construction generally should be carried out. To minimise the risk of expensive and time-consuming abortive work by the contractor, checks on materials and construction should be timed to take place soon after each trade has started on site – not as they are about to finish.

The list of essential items to inspect should include:

- fundamental items such as the setting out of the building on the site, and the setting out of walls and columns within the building
- details of particular aesthetic importance such as masonry bonding, and the setting out of light fittings and tiling
- parts of the building to which access will later be impossible, such as high-level external wall cladding, below-ground drainage, and screeds
- technically important elements that would be particularly expensive to rectify, such as reinforced concrete foundations, mortar strength, and the presence in the correct places of damp-proof membranes, vapour control layers, air leakage barriers, insulation, wall ties and damp-proof courses
- details that may cause danger to those occupying or maintaining the completed building, or the public – such as electrical cross-bonding, flues to gas appliances, and the fixing of high-level copings, masonry or cladding
- temporary protection of existing features, and of completed work
- innovative or 'risky' details and forms of construction
- tests, such as drainage tests, at which the architect should be present
- ordering of long lead-in materials and components – which may include structural steelwork, masonry units, metal windows, joinery and kitchen units, stone and proprietary worktop materials, and specialist finishes.

The contents and length of the list will obviously depend upon the nature of the job, and can be derived only from a thorough knowledge or careful study of the contract drawings and specification.

It is likely that the inspection plan will have to be updated as building work proceeds, but records of the original (and all subsequent versions) are evidence that the architect has done his or her job properly, and should be kept in the job files in case they are needed in defence against a claim of negligence.

4.2 Categories of inspection

The architect's inspections will fall into three categories:

- predictive inspections
- periodic inspections
- spot checks.

Predictive inspections are inspections planned in advance to ensure that important hidden elements of construction are inspected before they are covered up.

The architect will need to liaise closely with the contractor to ensure that inspections are made before items to be inspected are covered up as the work progresses. The contract may require – or it may in any case be worthwhile to ask – the contractor to give notice before such covering up is to take place. To be useful, notice should be given at least 24 hours before covering up.

Periodic inspections are inspections carried out at regular intervals to:

- check work executed since the previous visit (so far as it has not been covered up, or is to be opened up for inspection)
- monitor progress in relation to:
  - the contractor's programme
  - the date for completion
  - any other contractual dates
- determine whether or not progress is regular.

The frequency of periodic visits should depend entirely upon the requirements of the job, and could vary from every day for a very fast-track job with lots going on, to once a fortnight for a large, repetitive job. The frequency may also vary from one stage of a job to another.

Spot checks are occasional unannounced visits or tests to discourage the contractor from lapsing into poorer practices between the architect's regular visits. They should take place whenever particularly important work is in progress, when the architect has suspicions concerning the contractor's work, perhaps when a new trade has just started on site, or simply when the architect thinks such an inspection is due. Spot checking could range from arranging laboratory tests to establish the crushing strength of concrete or brickwork, to instructing the dismantling of a small area of fenestration or cladding to inspect the fixings. Adequate allowance for such tests and inspections should of course be included in the tender documents.

4.3 Importance of the contract documents

The contract documents are the single most important source of guidance to the inspecting architect. The contract documents define:

- the inspecting architect's authority and power
- the inspecting architect's duties under the contract
- the standards of workmanship, quality of materials, and quantity of work to be provided by the contractor.

Any attempts to vary the above during the course of construction will almost certainly affect timing or costs – or both. The inspecting architect must therefore have a thorough knowledge of the contract documents before carrying out site inspections.

As for the building work itself the drawings and specification are of most importance. If the architect carrying out the inspection has prepared the documents him or herself, as is often the case on a smaller job, he or she will be starting pre-inspection preparation with an advantage. However, even on the smallest of projects months can go by between completing and issuing production information when tenders are invited, and actually starting on site – during which time easily overlooked but vital features of details, setting out, or other important matters can be forgotten. It is therefore essential that the architect, even if he or she has prepared all the production information, refreshes his or her memory on the subject before building work starts.

If the architect carrying out the inspection has not prepared the relevant production information it will be impossible to carry out inspection duties effectively unless the time has been taken to study and become thoroughly familiar with the contract documents. On a large job it will not be possible for a single person to hold in mind at any one time every detail of the job. It will therefore be necessary to study in depth the information relevant to each stage of the job as the work proceeds.

4.4 Time monitoring

Before work starts on site the architect should assess the resources required to carry out inspection duties and their associated costs. It is not uncommon – often for perfectly valid reasons – for a significant amount of production information to remain to be prepared after the contractor has started on site. Such a situation can put an enormous amount of pressure on the architect in terms of time, resources and stress, and can thereby lead to the architect neglecting to monitor fees, or underestimating the costs of completing the job. The risk of running out of fees before the job is finished is thereby increased. Such a situation demands special care, and should be avoided if possible.

4.5 Advising the client

It is important before work starts on site that the architect reminds the client of the architect's responsibilities, and of the limits of the architect's powers, and that it is the duty of the contractor rather than the architect to supervise the work in progress. The client should be encouraged to attend or be represented at site progress meetings, and otherwise take an active interest in the work – which in itself will help to improve quality. The client should of course also be advised to make arrangements with the contractor before visiting site, and be reminded not to give instructions directly to the contractor.

4.6 Briefing site inspectors

Before briefing the clerk of works or other site inspectors, the architect should remember that he or she will retain responsibility for checking important elements of construction and for carrying out spot checks of the construction generally. If the architect is to rely on site inspectors to carry out some of the checks it is essential that they be briefed very carefully indeed. In all cases site inspectors should be reminded that their role is to look, and to report to the architect what they see.

Clerks of works are usually ex-tradesmen, and it should be possible to rely upon them to notice when normal standards of workmanship in traditional construction are not being achieved. Nevertheless, on all but the very simplest of jobs it will be necessary to brief site inspectors such that they fully understand and appreciate:

- the extent and limitation of their powers under the contract
- their duties
- the architect's and other consultants' construction details

- the reasons for and the importance of specific construction requirements
- unusual and special features of the design, and their importance
- the need to speak to the architect if in doubt about details, apparent discrepancies or any other matters.

The architect must – above all else – have clearly in mind the parts of the building that must be inspected at first hand, and the parts that could or should be left to others. Items of aesthetic and technical importance should always be inspected by the architect. Technically complex elements – which nowadays could even include such 'traditional' construction as a fireplace – should also always be inspected by the architect, consultant engineer or other suitably qualified person. If such items are repetitive, the prototype or first to be built should be inspected by the architect or engineer, who should continue to make spot checks at suitable intervals.

Before briefing site staff the architect should refer to the guidance in the latest edition of the *Architect's Job Book* published by the RIBA. If appropriate, site inspectors should spend some time in the architect's office before work starts.

4.7 Initial project team meeting

The initial project team meeting (or 'pre-contract' or 'pre-start' meeting) is important for many reasons. As far as the architect inspecting the works is concerned it is essential that during the meeting:

- the architect's and other consultants' staff likely to be involved in inspection, site inspectors, the contractor's manager and person in charge on site, and preferably a representative of the client are all present
- the respective inspection roles of the architect and site inspectors are clearly explained
- the contractor is made aware of particular details of construction that the architect wishes to see before they are covered up, and of tests that the architect wishes to witness
- it is stressed that, under the contract, supervision and quality control are the responsibility of the contractor
- the contractor be asked to confirm his proposed arrangements for ensuring adequate supervision and quality control
- the contractor hands over a copy of his contract programme
- the architect's and other consultants' programmes for the preparation of necessary further production information are discussed
- work by nominated or named subcontractors, or other specialist contractors, and the arrangement of separate meetings with such subcontractors or

specialist contractors as necessary to make clear the allocation of responsibilities for management, coordination, quality control and inspection are all discussed
- the contractor is reminded of his duty to understand the provisions within the design for means of escape and other fire safety measures, with a view to ensuring that such provisions are not compromised during construction
- the contractor hands over a copy of his health and safety plan
- special health and safety, temporary protection (of both internal, external and landscape items), and security requirements are discussed and understood as appropriate
- terms of party wall agreements are discussed and understood
- if the works involve a listed building the implications for new and existing structure, features, finishes, fittings and fabric generally are confirmed and understood
- it is pointed out that, under the contract, the architect has no duty to prepare 'snagging' lists towards the end of the job, and that any such lists compiled by anyone involved in the project should not be considered by the contractor to be exhaustive or definitive
- it is emphasised that practical completion is a matter for the architect's opinion alone, and does not depend upon whether or not the employer has taken possession of the works, or whether or not the contractor thinks he has attended to all items on particular 'snagging' lists.

If the above points have been raised and minuted, a lot of potential argument during and at the end of the contract can be avoided.

It is also a good idea to give the contractor, for displaying in the site office, some presentation material showing what the building will look like when it is finished. A visible goal will help to achieve a finished product of higher quality by encouraging and giving a greater sense of purpose to the construction team.

4.8 The contractor's programme

Under most building contracts the contractor's programme is not a contract document. The contractor cannot therefore usually be required to carry out the works exactly as shown on the programme. However, the programme is essential, both as a guide by which to monitor progress and as a basis on which to assess any claims for an extension of time – so the contractor should not start on site before issuing a copy to the architect. The architect should examine the programme very carefully and discuss any queries with the contractor. The architect's comments should be recorded in writing, but care should be taken to ensure they are not construed as instructions. Under no circumstances should the programme be approved by the architect.

Things to look out for include:

- dates shown incorrectly
- impracticable sequencing (such as electrical second fixing at the same time and in the same place as plastering)
- activities that are missing
- clearly inadequate time allowances
- inadequate allowance for 'hidden' activities (such as drying-out of plaster or screeds)
- subcontract programmes and lead-in times for materials and components not integrated into the overall programme
- inadequate allowance made for contingencies (such as, for example, work that may be required in connection with existing services or structure below ground).

The same checks should be made if the programme is revised and reissued at any stage during the job.

Sometimes, in order to please impatient clients, contractors are asked to carry out work unrealistically quickly. Contractors may agree to such timescales, confident that grounds for claims for extensions of time and for making exorbitant loss and expense payments will arise – or can be cooked up. The architect should avoid such situations.

4.9 First meeting with the site agent

An excellent start can be made by immediately before or during the early stages of the contract holding a meeting with the site agent to explain the drawings and specification. The meeting should be informal, attended only by the architect and the agent, and should be separate from the formal initial contract or 'pre-start' meeting.

The site agent is without doubt the most important person on the site, and the success of the job very much depends upon the relationship between the site agent and the job architect. An early, businesslike meeting between the two will create a good impression, and help to establish a productive relationship.

Once the job is under way it is inevitable – even with the best of contractors, and with all production information available and complete – that work not in accordance with the drawings and specification will be found. Apart from sheer carelessness – which it is impossible to eliminate entirely – the main causes will be that the architect's details are simply not understood; that specific requirements are, despite the best possible intentions, overlooked; or that the reasons for certain construction requirements are not appreciated – leading to their being misinterpreted or simply ignored in favour of alternatives that are more expedient to the contractor or tradesman.

The initial meeting with the site agent should help to minimise the occurrence of such faulty work by:

- providing an opportunity for the architect to emphasise important construction details
- allowing for the resolution of any immediate queries that the contractor may have
- consequently bringing to light any obvious gaps in the production information.

The meeting is also a good opportunity for the architect to remind the site agent of the particular details of construction that the architect wishes to see before they are covered up, and of tests that the architect wishes to witness.

During such meetings the architect should bear in mind that, contrary to the contractor's assurances that for weeks before starting on site the site agent has been assiduously studying the contract documents, it is possible that the agent left his previous job on the Friday before starting on site on the Monday, and was given the drawings for his new job only in time for them to be opened briefly during the weekend.

5 While work is in progress

5.1 The timing of inspections

Before work started the architect should have prepared a plan, based on the contractor's programme, showing when inspections are to be made. The plan should be updated as necessary to accord with actual progress on site. Inspections should be carried out in accordance with the plan.

Special attention will be required to the timing of predictive inspections and visits to witness the carrying-out of tests. It will be necessary to keep in close touch with the contractor to avoid visits being made too late, or having to instruct the postponement of work that would otherwise cover up work to be inspected. It may have been agreed, or the contract may require, that the contractor give notice before particular items of work are ready for inspection (although failure of the contractor to give such notice would not excuse the architect from inspecting work before it is covered up).

The timing of periodic visits is not so critical; they need simply take place regularly, except that their frequency should be increased or decreased as necessary to suit the requirements of each stage of the job. Spot checks need be made as and when appropriate. Inspections should be timed so that one is always made shortly before issuing certificates for payment.

5.2 Use of checklists

Before each visit to site it is essential that the architect reviews the drawings, specification, and inspection plan in relation to the stage that work has reached, and prepares a list of specific items to be checked. Checklists taken from a book or article, or an office standard checklist, will include the more obvious and commonly found defects and can be useful, but such lists must be adapted to suit the particular project, its priorities, the contractor, the way the job has gone so far, and the architect's experience of similar jobs in the past.

It is good practice to set out each checklist on a standard site inspection form that provides plenty of space for comments – to be added during, or immediately after, the architect's visit to site – against or below each item to be inspected (see Figure 5.1). The architect must not forget to take the checklist with him or her when inspecting work, as the numerous distractions on site could otherwise cause items from the list to be forgotten.

SITE VISIT REPORT	
Job:	Date:
Job no:	Visit by:
No. of visits scheduled:	Visit no:
Purpose of visit:	Page of

Items to be checked   **Observations**   **Actions**

Storage and protection
Health and safety
Contractor's supervision
Trades and no. of operatives on site
Progress in relation to programme
Delays and causes
Information required

Checked   **Records**
Samples: Photos:
Tests: Video:
Vouchers: Other:
Contractor's records

Figure 5.1 Site Inspection form

Where the contract documents refer to British Standards, Agrément Certificates, industry codes of practice or similar publications it is advisable to check their contents before leaving the office – or to ask the contractor to make them readily available on site.

5.3 Priorities for inspection

It is never possible, even with a full-time site architect and clerk of works, to check every item of work carried out. It is therefore necessary to ensure that inspection efforts are concentrated so as to make the most effective use of time and resources. Items given priority for inspection should include:

- setting out of the building on the site
- protective coatings to structural steelwork
- cavity wall construction
- special tolerances
- roof details
- details designed to prevent the penetration of water or damp
- setting out of partitions, and setting out and dimensions of internal and external openings
- storage and protection of masonry units, cladding components, windows, doors and joinery units
- setting out and fixing of services before plastering
- fire-stopping at services penetrations, and other elements of fire-resisting construction
- screeds
- fixing of glazing in frames
- the work of any trade that has started since the architect's last visit to site
- boundary conditions with respect to the terms of party wall agreements, and statutory and other consents and approvals
- protection of existing building or landscape elements to be retained
- other elements of work shown on the architect's inspection plan.

The list will vary from job to job.

5.4 Once on site

When on site to carry out inspections there are a number of general principles that the architect must bear in mind:

- The architect must always report to the contractor's person in charge immediately upon arrival, and must not visit the site outside working hours when the contractor's staff are absent.

- The architect must actually inspect. Inspection means looking at specific elements of work to find out whether or not they comply with the requirements of the building contract. Inspection does not mean making assumptions based on general impressions gained by wandering around the site.
- The specific elements inspected will either comply or not comply; it is not part of the architect's job to decide whether or not work probably complies.
- The function of inspecting should not be confused with the functions of visiting site to attend meetings or to deal with queries. (Once on site it is common for the architect to find that all his or her time is being spent answering large numbers of questions from the contractor and trades foremen, or discussing particularly tricky problems that have arisen since the architect's previous visit. Although such matters are obviously important, inspecting should be kept separate from other activities, and sufficient time must be dedicated to it.)
- It is vital that the architect distances him or herself from the turmoil and looks carefully at the particular elements of the building construction that he or she has come to site to inspect. This is best done alone, without distraction from contractor, client or anyone else.
- Inspection must be done thoughtfully and methodically, comparing the building work on site with the construction drawings and specification. The architect must think through the implications of what has been built and what has not been built, and bear in mind the principles behind particular details being inspected.
- It can be helpful to follow a set route that starts with the parts of the building that are the most advanced, takes in all parts of the site in sequence, and finishes with the least advanced parts of the building. With a new building it will probably mean starting at the bottom and working up, but with a refurbishment or conversion it may mean starting at the roof and working down. Such a method assists the monitoring of progress and minimises the risk of inadvertently neglecting to inspect parts of the site.
- The storage of materials and the protection of finished work can be as important as work in progress.
- Packaging – either of materials stored ready for use, or of those discarded after use – can be helpful in determining whether or not the right materials have been used.
- Watching operations in progress may bring obvious errors of workmanship to the architect's attention, but the architect should be aware that the work being carried out may not be representative of what goes on during his or her absence.
- Giving a new trade special attention will help to set standards, resolve initial difficulties, and spot mistakes before they are repeated.
- The architect must inspect everything on the checklist, but also take time to step back and look at things in general, taking a critical overview of the site as a whole. He or she must be alert to the possibility of discovering the unexpected.

- It should not be forgotten that the authority and powers of the architect on site extend only as far as the contract documents permit. Any attempts by the architect on site to extend his or her powers or impose requirements not included within the documents will almost inevitably lead to the incurring of delays, or additional costs to the client – or both.
- The architect should be firm with the contractor about what should be done, but should resist any temptation to tell the contractor how to do it. Instructing the contractor on how work should be carried out could leave the architect liable for additional costs, damage to property, or in the worst cases injury or death.
- The architect should not give instructions directly to operatives – even if they are doing things wrong – unless health or safety are immediately at risk. To do so can lead to misunderstandings and undermine site management.
- The architect should not be put off from inspecting work because access is difficult or unsafe, but should insist that adequate safe access be provided. Lack of access is not an excuse for over-certifying.
- When preparing valuations the quantity surveyor will measure all work that is evident. However, it is not part of the quantity surveyor's job to take a view on whether or not work complies with the requirements of the contract documents. The architect must therefore ensure that the quantity surveyor is informed of all defective work, and that payment is certified only in respect of work properly executed.

5.5 Contractual provisions

Building contracts can give to the architect specific powers intended to facilitate checks on materials and workmanship. Clause 8.2.1 of the JCT's Standard Form of Building Contract 1998 Edition (JCT 98), for example, provides that:

> *The Contractor shall upon the request of the Architect provide him with vouchers to prove that the materials and goods comply with [the requirements of the contract documents].*

Where materials are used extensively or where their role is critical test certificates or other 'vouchers' should always be requested.

Clause 8.3 of JCT 98 also provides that:

> *The Architect may issue instructions requiring the Contractor to open up for inspection any work covered up or to arrange for or carry out any test of any materials or goods (whether or not already incorporated in the Works) or of any executed work, and the cost of such opening up or testing (together with the cost of*

making good in consequence thereof) shall be added to the Contract Sum unless provided for in the Contract Bills or unless the inspection or test shows that the materials, goods or work are not in accordance with this Contract.

Clause 3.12 of the JCT's Intermediate Form of Building Contract (IFC 98) contains similar provisions.

Such a provision is invaluable, as some defective work can be detected only by opening up. It is therefore worth including in the contract documents, for any but the smallest of jobs, a provision for opening up and testing, so that once on site the architect can instruct any necessary opening up without adding to the contract sum.

The architect's powers to request opening up or testing must be used selectively, but they must be used. Elements of work that it might be appropriate to inspect by opening up or testing – or both – could include:

- hidden damp-proofing details where the contractor may have been tempted into doing a quick shoddy job rather than a proper one
- repetitive elements, where standards may have been allowed to slip
- depths and mix of screeds
- depths and quality of asphalt
- crushing strength and porosity of masonry units
- wall construction and mortar mix
- moisture content of timber
- watertightness of plumbing and drainage
- airtightness of flues
- fixing of windows, cladding or other components.

It is important that the architect instructs the opening up or testing before following trades – for example floor finishes on screeds – commence.

5.6 Storage and protection

Components and materials can become stained, chipped, scratched, dented or otherwise damaged if they are badly stored or protected. There may not be time to reorder damaged stored components, or it may be physically impracticable to replace components or materials damaged after they have been incorporated into the works. Making good to damaged work is never entirely satisfactory, and the need for it should be avoided. Openings being used for access during construction are especially vulnerable, as is finished work next to routes through or around the building.

Door jambs and heads, door leaves, window sills, stair treads and strings, nosings, balustrades, floor finishes, sanitary ware, and any form of pre-finished component from a cladding panel to a light fitting are likely to require careful attention. The architect should see that they are properly protected both before and after they are fixed, that they are not fixed too early, and that where possible the contractor avoids potential damage to vulnerable parts of the building by using alternative or temporary means of access. Similarly the contractor should not be using WCs that he has not been expressly permitted to use, should not be washing paintbrushes in basins or sinks, and should not be using vanity unit tops as workbenches. It may be best to remove fitted components temporarily, protect them in a secure part of the site, and refit them shortly before completion.

Equal care should be taken of existing features to be retained. If work involves a listed building the consequences of failing properly to protect existing features, finishes, fittings and building fabric generally can be particularly onerous. Externally, the architect must pay special attention to the protection of trees to be preserved.

The architect should also be aware, particularly when certifying payment, that the ownership of materials stored on site may not have passed from the supplier to the contractor.

5.7 Inspecting work off site

There are occasions when it is of benefit for the architect to inspect elements of work off site. Visiting a workshop during the making of an important piece of metalwork or joinery can help to ensure that standards of workmanship are adequate and that details of construction have been understood. Visiting a quarry while work is in progress can ensure the consistency of stone supplied to site.

Visits should be timed to be neither too early for anything worthwhile to be seen, nor too late to avoid risking significant abortive work if things are not as they should be. Such visits are often provided for in the contract documents – and the visits should be made. In the absence of express provisions the contractor's agreement should be obtained. Although inspecting work off site may be important for the purpose of quality control it is likely that ownership of the work will not have been passed to the contractor. The architect should therefore be extremely reluctant to certify payment to the contractor for such work.

5.8 The contractor's quality control procedures

As well as the architect inspecting the work him or herself it is important to check that the contractor's own quality control procedures are being properly implemented. The architect's client is paying for the contractor to control quality, and the contractor should not be relying on the architect's inspections instead.

The procedures to be put into effect by the contractor will have been determined by the specific requirements of the contract documents, and the proposals agreed during the initial project meeting. The contractor may have been required to provide a method statement.

Essential is the permanent presence on site of a competent person who can and does supervise the work on a day-to-day basis. Such a provision should be an express requirement of all building contracts. The failure of a contractor to satisfy such a requirement should be treated by the architect extremely seriously. Trades foremen must also be competent, and present on site as necessary to ensure that prescribed standards of workmanship and progress are achieved. The operatives themselves must be appropriately skilled and experienced for the type of work they are doing.

It is of equal importance that management and workers on site have the correct production information, and are using it. The architect should check that the right information is being used, and may also – with the contractor's permission – help to avoid defects by explaining to site management the aesthetic, practical and technical reasoning behind the design and its details. This may be done during an initial informal meeting with the site agent, or stage by stage with key personnel as the job progresses – or a combination of both.

The contractor is often required by specifications to have on site copies of all standards and codes of practice to which the specification refers. On a small domestic job such a requirement is probably unrealistic and unlikely to be satisfied, but on larger jobs the architect may insist that standards are obtained and that the contractor refers to them.

Management environments within which good-quality building work is produced are likely to feature:

- meetings between main contractor and subcontractors well before subcontractors are due to start on site (the purpose of such meetings being to ensure that before the subcontractor starts work he knows what to do and with whom to raise queries, that the main contractor knows what to

provide in terms of access, builder's work and materials, and that when the subcontractor arrives on site he can have ahead of him a clear uninterrupted run of work)
- managers who clearly understand their position and function
- site managers who are competent, and sufficient in number to allow time to coordinate, supervise and check the work of all trades
- good welfare facilities for operatives
- regular minuted meetings to solve problems, with contributions welcomed from a wide range of people.

5.9 Contractor's records

On all but the smallest of jobs the contract documents should require the contractor to keep on site records of the contractor's quality control activities, and matters affecting progress.

Records of quality control activities should clearly identify the subject of each individual record by element, item, batch, lot and location in the works as applicable. Such records should include:

- dates, details and results of the contractor's own inspections and tests
- documentary evidence of the quality of materials and goods being incorporated into the work
- details and extent of any work found not in accordance with the contract documents
- details of corrective action taken.

Records of matters affecting progress should include:

- a continuously updated record of the progress of the works, shown on a copy of the programme
- daily records of the number and description of all main and subcontractor's personnel on site, including site management
- daily records of the number, type and capacity of all mechanical and power-operated plant on site
- maximum and minimum temperatures in each 24 hour period
- the number of hours per day in which work is prevented by adverse weather.

The architect should not only check that records as required by the contract documents are being kept, but should also carry out occasional spot checks of their accuracy.

5.10 What should the architect do if defective work is found?

If the architect finds work on site that is not in accordance with the contract documents it should be immediately reported to the contractor's site management. The architect should not directly ask the operatives carrying out the work either to stop the work, or to do something different than they are doing. If the architect has questions about what is being or has been done they should be asked through the contractor's site manager – or at least in the manager's presence.

Many contracts include provisions for use in the event of defective work being found. Clause 8.4 of JCT 80 provides that:

If any work, materials or goods are not in accordance with this Contract the Architect, without prejudice to the generality of his powers, may:

1 notwithstanding the power of the Architect under clause 8.4.2 ... issue instructions in regard to the removal from the site of all or any of such work, materials or goods; and/or

2 after consultation with the Contractor ... and with the agreement of the Employer, allow all or any of such work, materials or goods to remain and confirm this in writing to the Contractor (which shall not be construed as a Variation) and where so allowed and confirmed an appropriate deduction shall be made in the adjustment of the Contract Sum; and/or

3 after consultation with the Contractor ... issue such instructions requiring a Variation as are reasonably necessary as a consequence of such [instructions issued under 1 and 2 above] and to the extent that such instructions are so necessary ... no addition to the Contract Sum shall be made and no extension of time shall be given; and/or

4 having had due regard to the Code of Practice appended to these Conditions ... issue such instructions under clause 8.3 to open up for inspection or to test as are reasonable in all the circumstances to establish to the reasonable satisfaction of the Architect the likelihood or extent, as appropriate to the circumstances, of any further similar non-compliance. To the extent that such instructions are so reasonable, whatever the results of the opening up for inspection or test ... no addition to the Contract Sum shall be made ...

There are a number of points worth noting in connection with the clause, as follows:

- The architect is not restricted to choosing only one of the four options – any combination of more than one may also be employed.

- There is no provision for the architect to instruct that the defect be corrected: it can only be, in whole or part, removed from site, accepted, or accommodated by means of a variation.
- Accepting or accommodating defects under such a clause should be done only at the request of the architect's client, and even then the architect should first consider responsibilities to owners, occupiers and third parties, and act accordingly.
- The provisions for opening up and tests under clause 8.4.4 are different from the provisions under clause 8.3.

Under clause 8.3 the architect can instruct the opening up or testing of work without evidence of a defect. However, the employer will pay for the opening up or testing and consequent making good unless work not in accordance with the contract is revealed.

In contrast, clause 8.4.4 does not entitle the architect to instruct any opening up or testing unless defective work has first been found (perhaps, but by no means necessarily, as a result of previous opening up or testing under clause 8.3). In such a case, whatever the results of the opening up or testing, and provided the architect's instructions to open up or test were reasonable, the employer will not have to pay the costs (although, if the opening up or testing reveals no further defective work, the architect will have to grant an extension of time for any associated delay to completion).

The provisions in IFC 98 are somewhat different. If defective work is discovered the contractor is required – without instruction or request from the architect – to comply with clause 3.13.1, which states:

> *If a failure of work or of materials or goods to be in accordance with this Contract is discovered during the carrying-out of the Works, the Contractor upon such discovery shall state in writing to the Architect ... the action which the Contractor will immediately take at no cost to the Employer to establish that there is no similar failure in work already executed or materials or goods already supplied (whether or not already incorporated in the Works). If the Architect ... :*
> - *has not received such statement within 7 days of such discovery, or*
> - *is not satisfied with the action proposed by the Contractor, or*
> - *because of considerations of safety or statutory obligations, is unable to wait for the written proposals for the action from the Contractor,*
>
> *the Architect ... may issue instructions requiring the Contractor at no cost to the Employer to open up for inspection any work covered up or to arrange for or carry out any test of any materials or goods (whether or not already incorporated in the Works) or any executed work to establish that there is no similar failure and to make good in consequence thereof.*
> *The Contractor shall forthwith comply with any instruction under this clause ...*

It is interesting to note:

- Neither the contractor nor the architect can exercise discretion as to whether or not the contractor must submit his proposals to the architect. (However, if the contractor objects to an instruction issued by the architect under clause 3.13.1 he is given the right by clause 3.13.2 to appeal in writing to the architect within 10 days of the date of issue of the instruction. If, within seven days of the date of issue of the contractor's appeal, the architect has not withdrawn or modified the instruction to the contractor's satisfaction, the matter is left to be decided under the dispute resolution procedure applicable under the contract.)
- The contractor is theoretically required to write to the architect after every discovery of a defect, which will inevitably include a large number of minor defects that can be very simply put right. The architect and the contractor should therefore use common sense in operating the clause and, if practicable, during the initial project meeting agree how the clause is to be used.
- The architect must give an extension of time for any delays to completion caused by opening up or testing under clause 3.13.1, unless defects are found.

(Clauses 3.14.1 and 3.14.2 of IFC 98 give the architect powers roughly equivalent to those given by clauses 8.4.1–8.4.3 of JCT 98.)

The architect must exercise care in operating clauses such as those referred to above so as to avoid inadvertently adding to the contract sum. In particular, the architect must be wary of using the provisions of the clauses to improve dubious details. For example, the architect may instruct the opening up of a cavity wall and find there are fewer wall ties than specified. If the architect then instructs the contractor to rebuild the wall with the specified number of wall ties the contractor must do so at his own expense. However, if the architect instructs the contractor to rebuild the wall with more wall ties than were specified at the time the wall was first built – or with higher-quality wall ties – it is likely that the contractor will have a good case for claiming that more than the cost of the additional or higher-quality wall ties be added to the contract sum: the instruction will effectively have become a variation.

Whatever the contractual provisions it is important that the architect operates them strictly, firmly, and promptly. The removal from site of defective work and its replacement with work in accordance with the contract documents can cause significant delay, and can be expensive for the contractor. The architect can therefore be found under considerable pressure from the contractor – or even the client – to accept work that is inadequate or botched. The architect has no authority to accept such work, and to do so would leave the architect open to accusations from the client and possibly from third parties. The safest course is to have the work removed and rebuilt correctly. The worst thing to do is endlessly to prevaricate.

However, if for any reason it appears preferable to accept work not in accordance with the contract documents, the potential consequences must be given the most careful consideration before a firm decision is made. Apparently harmless changes during the course of building operations can have far-reaching and serious consequences that are difficult to predict while in the middle of a stressful construction programme when important reasoning behind the original details may not immediately return to mind. Matters to which special thought should be given include:

- function
- technical performance
- finished appearance
- durability
- effects on safety, fire protection and means of escape.

If the client disagrees with the architect's conclusions the architect should write to the client recording the facts and disclaiming responsibility, taking into account duties to third parties.

Details of defective work, whether found by observation or by opening up, should be recorded by the architect in writing and – except perhaps for isolated minor items – by photograph.

Finally, the architect should remember that a defect in construction is a failure of the contractor's quality control system. Following the discovery of defects the architect should insist that the contractor improves his supervision of the work.

5.11 Health and safety

Site safety is the responsibility of the contractor. The architect has no duty either to devise or to approve safe methods of work, and is neither trained nor paid to do so. To avoid contributing to or being held liable for an accident the architect should be extremely careful to avoid involvement in such matters. However, if the contract documents require the contractor to submit a health and safety plan or method statement the architect should carry out checks on its implementation.

If, while carrying out the architect's inspection duties, the architect thinks that any work or other activities taking place on site appear unsafe, he or she should immediately raise the matter with the contractor. A failure to do so may lead to an accusation of tacit approval. If the architect remains concerned, and the contractor refuses to do anything about it, the architect should immediately contact the Health and Safety Executive. If third party property or members of the public may be in danger, the architect should as appropriate call the local building control officer or even the police. Failure by the architect to act in such circumstances could lead to him or her being held liable. If a contractor does not cooperate in connection with health and safety matters, payment for any affected work should not be certified, on the basis that it is not being properly executed.

Obvious things to look out for include safety rails and toe boards fitted to scaffolding, proper fixing of ladders, adequate hoardings and screens to protect neighbours and the public, and safe access generally in and out of and around the site. The architect should refer for further guidance to *Health and Safety in Construction* published by the HSE.

All action taken in connection with health and safety should be recorded in writing.

5.12 Site security

Site security is also a matter for the contractor. However, as with safety, if the architect thinks something is not right he or she should express such concerns to the contractor. Neglecting to do so may be interpreted as tacit approval. Security requirements should have been clearly set out in the tender documents, but it may be necessary to increase security provisions beyond those specifically required at tender stage. The architect should remain alert to such matters. A written record of all action taken should be made and kept.

5.13 Welfare facilities

The quality of welfare facilities provided by the contractor for operatives plays an essential role in:

- attracting the best subcontractors and individual tradesmen
- keeping up the morale of those on site
- thereby increasing the quality of the finished job.

The contract documents should therefore require the contractor to provide and maintain, in fully working and clean condition, welfare facilities of a high standard. The employer under the building contract will consequently be paying for such provision and maintenance, and the architect should make sure that the contractor provides them.

5.14 Site tidiness

Tidiness of the site is a matter of efficiency and safety. A site that is strewn with randomly stored and inaccessible materials, and which is littered with debris and rubbish, is not conducive to the production of high-quality work, and is simply not safe. Such a site should not be permitted by the contract documents and should not be tolerated by the architect.

5.15 Monitoring progress

An architect administering a contract needs to monitor:

- actual progress being made on site
- events that may have a bearing on potential claims.

The purpose of such monitoring is to keep the client informed of likely delays and additional costs. It is best done by:

- comparing actual progress on site with the contractor's programme
- checking the ordering and fabrication of long lead-in materials and components
- asking the contractor for advice on matters that may cause or are causing delays or disruption
- keeping an eye out for potential problems.

Monitoring should be done regularly and frequently to ensure that potential delays are discovered at the earliest opportunity. The architect must not simply rely on the marked-up programme in the contractor's office or on assurances from the contractor that progress is as it should be, but must carry out his or her own checks and assess the situation for him or herself.

Matters that commonly cause delay include:

- late production information
- inadequate forward planning by the contractor
- components or materials ordered too late, or subject to increased delivery periods
- scaffolding, or other means of access around or into the site not being available on time
- subcontractors not arriving on site when due, being prevented by other commitments from returning to site promptly after postponing visits because of earlier delays, or not proceeding regularly and diligently
- insufficient or poorly skilled people deployed on site

- problems arising from discrepancies in production information, or the impracticability of details
- the need to make good defective work
- variations
- statutory consents not obtained, or conditions not satisfied sufficiently in advance of affected work
- work by statutory undertakers not completed when required
- delays by the client in supplying materials or components
- delays or disruption caused by the client's specialist contractors
- prolonged frost, severe rain, or snow.

The architect should be alert to the potential occurrence of any such events. If it appears that any of the above events is likely to occur, the architect should promptly take up the matter with the contractor with a view to evasive action being taken. The architect must report to the client as necessary.

5.16 Daywork sheets

An inspecting architect should not sign daywork sheets unless he or she has authority to do so, has been given adequate notice of the commencement of the work, has been given a reasonable opportunity to verify the accuracy of the facts recorded, and has been presented with the sheets no later than the end of the week after the work was carried out.

The contractor should also be reminded that the signing of daywork sheets does not:

- in itself constitute an instruction
- confirm acceptance of cost implications
- mean that the work will necessarily be valued on a daywork basis.

On a large job it may be worth obtaining and using a rubber stamp worded with such a reminder.

5.17 The clerk of works and other site inspectors

The architect is responsible for the direction of the clerk of works and other site inspectors. The architect should therefore keep in close contact with them and, as work on site proceeds:

- explain priorities for inspection
- explain complex or important details, the reasoning behind them, and the intended results – from both the aesthetic and technical points of view

- ensure that site inspectors have adequate facilities, supplies of forms, etc
- ensure that procedures are followed and that records are completed correctly, returned on time, and properly filed (clear records of competent inspections by site staff could be invaluable in the event of a dispute)
- ensure that site inspectors are not exceeding the limits of their authority by advising or instructing the contractor, trades foremen or operatives – or by other means
- ensure, by the quality of the architect's and other consultants' own production information, that there is no need for site inspectors to waste time interpreting and completing it for the contractor's benefit
- provide support and back-up to site inspectors, and ensure that they are not undermined (for example, by the architect being lenient to the contractor over a matter on which a site inspector had been firm).

The architect must remember that the duties of site inspectors are to check, inspect, and report – and no more. The purpose of site inspectors is to provide more frequent and detailed inspection services in addition to, not instead of, those to be provided by the architect. Unless the architect's appointment contains express provisions to the contrary, site inspectors' services are not a substitute for the architect's inspection services.

Therefore the architect must not delegate to site inspectors duties that are the responsibility of the architect. Everything important must be inspected by the architect. Where a project includes a number of repeated elements the architect should inspect the first to be built, should periodically check elements subsequently built and, provided no defects are found, may otherwise delegate inspection to the clerk of works or other site inspector. When delegating inspection to a site inspector it is essential that the inspector be carefully instructed on what to look at, what to look for, and why.

Should the architect find that a site inspector is unreliable in any way the architect should immediately take up the matter with the site inspector's employer. If the situation is not improved it may be necessary for the architect to increase the length and frequency of his or her own inspections. Under no circumstances should an architect continue to delegate matters of any importance to a site inspector who the architect believes is unreliable or incompetent.

5.18 Other consultants

If acting as lead consultant the architect should ensure that other consultants perform their own inspection duties, notify the architect of any defective work found, and submit to the architect written inspection reports. To help avoid over-certification the architect must ensure that the quantity surveyor is kept fully updated and informed on all matters concerning defective work inspected by other consultants so that it may be excluded from the quantity surveyor's valuations.

5.19 Inspection records

During or immediately after each visit to site records must be made – and kept – of all items of work inspected and observations noted. (If records are completed while on site it is essential that writing be legible. Some architects use a tape recorder and, in addition to writing up notes when they return to the office, may keep the tape.) Although inspection notes need only be brief, they must be sufficiently comprehensive to enable others to understand them long after the visit has taken place. A good policy is to use a simple office standard form, such as that illustrated in the *Architect's Job Book*, which should be carefully completed and properly filed. Some architects issue copies of their site inspection records to the contractor and other parties as a reminder of actions to be carried out. Such a practice is not necessary and probably best avoided, but if followed it is essential that copies of site records issued as reminders are not confused with architect's instructions.

Photographs and video tapes are of invaluable help in recording defects and progress. Such records should be made regularly at appropriate intervals throughout the course of a job. Prints and tapes should of course be dated, labelled as necessary to identify the subject of each shot or tape, and filed with the written records of the visit during which they were taken. The rate of progress may best be recorded by annotating, hatching or otherwise marking up a drawing, or a copy of the contractor's programme – or both.

While completing records the architect should bear in mind their potential uses as:

- a reminder of any questions that need answering when the architect returns to the office
- a reminder of anything that the architect needs to re-inspect when next visiting site
- a reminder of work not properly executed of which the quantity surveyor should be advised so as to give him the opportunity of excluding it from his valuations
- a record of progress to assist in assessing potential claims for extension of time or reimbursement of loss and expense
- a record for use in justifying additional fees for necessary visits to site that at the time of the architect's appointment could not reasonably have been foreseen

- a vital record for use as evidence in defence against a claim of negligent inspection.

Matters to be recorded should include details of:

- work, samples, and stored materials inspected
- tests witnessed
- vouchers, contractor's records, and other paperwork checked
- defective work, and rectification required
- action taken to check previously executed work upon discovery of defects
- queries raised by the contractor
- contractor's notices of work being ready for inspection
- delays, including periods of time, items affected and causes
- the number and description of operatives and plant on site, with comments on the effectiveness of their deployment.

Other consultants should be asked to provide copies of their own site inspection records, which should be carefully filed by the architect.

In conclusion, it is worth noting two important points of principle:

- Records comprising a litany of unreasonable levels of defective work found on visit after visit may show that the architect has been very good at observing and recording, but that he or she is not doing all that should be done to achieve the raising of standards.
- In spite of the importance of records, it must not be forgotten that the architect's time is best spent actually inspecting the works, and should not be wasted on superfluous paperwork.

5.20 Completion

Inspecting work during the final stages of a job can be a frustrating experience for the architect. Theoretically, as the work nears completion, the contractor should himself carry out inspections as necessary to ensure that by the time the building is due to be handed over there are no defects in the work of finishing or other trades.

However, it is rare for a contractor to take it upon himself to carry out such inspections properly. It is more likely that a contractor will either before completion ask the architect for a 'snagging' list of everything that still needs doing, or simply present as complete to the architect work that close inspection reveals is far from ready. Such a situation can be made more frustrating if the architect's client is under urgent pressure to take possession.

Whatever the circumstances the architect must ensure that his or her own position remains strong. Under the standard forms of contract the contractor has no right to demand a snagging list from the architect. If the contractor asks the architect to prepare a snagging list the architect should refuse. The architect is not normally paid by the client to prepare such lists for the contractor, and to do so can waste a great deal of the architect's time. Furthermore, if the architect provides such a list, the contractor will be tempted to raise objections if, after the list has been issued, the architect asks the contractor to attend to matters that, for one reason or another, were not included in the issued list.

To avoid potential misunderstandings the architect must, at the initial project meeting, confirm to the contractor – and as the work approaches completion remind the contractor – that it is the contractor's job to ensure that work is properly completed and ready to hand over, and that the architect will not be providing snagging lists. A practical approach is for the architect to advise the contractor that if, when work is presented to the architect for final inspection, more than a certain number of defects are found, the architect will immediately stop inspecting and will not continue until the work has been properly completed. If it becomes necessary so to suspend an inspection the architect should not return to re-inspect until reasonably confident that the work has been properly completed – however long it takes. In the meantime the architect should ensure that the value of all outstanding and defective work remains uncertified, and that a practical completion certificate is not issued until at the earliest there are no visible defects and only a small number of minor items remain outstanding. The architect thus protects the client's interests, avoids having to type and issue long lists of defects, and gives the contractor every incentive to complete the job both properly and promptly.

The strategy is most effective when the employer under the building contract is able to resist taking possession until the contractor properly completes the work – even if the contractor is late. The employer is then able to continue holding all of the contractor's retention money until the contractor finishes the job properly, and may also be able to levy liquidated damages for late completion – thus giving the contractor every incentive to finish well. Even if the employer, by agreement with the contractor, decides to take possession in spite of the work not being properly finished, the architect should under no circumstances issue a certificate of practical completion until the work is properly completed, and should ensure that there remains enough money uncertified to account for all outstanding or defective work. In the worst cases it will be necessary for the architect to invoke the relevant contract provisions and arrange for another contractor to complete the work. Enough money should remain uncertified to pay the second contractor.

Although, during the final stages of work on site, the architect should not be providing the contractor with snagging lists, the architect should nevertheless continue to make records of inspections as has been done throughout the job. Such records may well include lists of defects or outstanding items of work – for the architect's or other inspector's own use as an aid to memory when monitoring progress. As the job approaches completion it is likely that the frequency of the architect's inspections will need to be increased to allow the architect to monitor progress properly and to keep the client fully informed.

5.21 At the end of the contract

If the contractor's performance has been good the architect should at an appropriate point after work has been completed write to the contractor to thank him for his efforts, to offer to act as a referee, and to acknowledge the achievements of any particularly deserving individuals within the contractor's organisation.

6 Practical matters

6.1 What should the architect wear on site?

The choice of clothing to wear on site should be determined largely by common sense. Footwear is particularly important. Shoes should be flat, strong and have tough, well-gripping soles. Leather soles are not suitable for climbing ladders or steeply sloping surfaces. Training shoes should not be worn. On some sites it will be necessary to wear Wellington boots or boots with steel-protected toe caps and soles. Such boots are normally provided by the contractor. If the architect is to visit site regularly he or she should make sure that the contractor obtains and provides boots of the right size.

Clothing generally should be smart but practical. The architect should not be deterred from inspecting properly for fear of spoiling a new suit, but neither should he or she appear scruffy. A good rule is not to appear less formally dressed than the site agent. A tough waterproof overcoat can protect more vulnerable clothes from most building site hazards.

A hard hat, normally provided by the contractor, should always be worn.

6.2 What else should the architect take on site?

When on site it is difficult to remember all the details of the parts of the building that are to be inspected. It should be possible to check details by reference to the contractor's drawings, but it is sometimes the case that they are not up to date, or that the relevant drawing has been removed from the site agent's clip and cannot quickly be found. It is therefore useful to take onto site copies of all drawings relating to the parts of the building to be inspected – either held together with bulldog clips and rolled up, or reduced to A4 or A3 size and contained in a plastic envelope. The architect can then be certain of having to hand all the production information needed, and can easily point out to the site agent any drawings missing from the contractor's site set (which the contractor should be told to rectify).

Other essential equipment includes a steel tape measure – the 5 m size is best, as when extended it is rigid enough to allow most vertical and horizontal dimensions to be checked without needing someone to hold the other end; and of course something to write on. A torch, a pocket spirit level, and a plumb bob can also be useful. A camera will be needed when photographs or video recordings are to be taken. If the architect will need to climb ladders or other potentially hazardous obstacles while carrying any of the above equipment it is essential to have either big enough pockets, or a bag that can be carried on the back so as to ensure that both hands are completely free for holding on.

6.3 How should the architect conduct himself or herself on site?

The success of an architect inspecting work in progress will depend very much upon the relationships between the architect and the people with whom he or she comes into contact on site. Inevitably relationships between people are determined in part by the personalities involved, and every architect will develop a personal style for dealing with people on site. However, productive relationships are more easily maintained if a number of basic general guidelines are followed.

The architect will achieve most by remaining at all times polite, courteous and tactful. At times the architect will need to be firm, but will never get the best from the contractor and subcontractors unless he or she is also approachable, encouraging, appreciative and helpful. The architect should take the trouble to learn the names of those with whom he or she comes into contact. Humour can be invaluable.

It is most important that basic etiquette relating to safety is followed: the architect should never visit site when the contractor is absent, should always report to the contractor's person in charge immediately upon arrival, and should not walk onto the site without appropriate protective clothing such as a hard hat or boots. It would be potentially dangerous and extremely ill-mannered for the architect simply to arrive on site and start wandering around without first letting the contractor know where he or she will be and what he or she will be doing. The contractor should also be advised when the architect is about to leave the site. Failure to follow such basic rules will cause the architect to appear unprofessional, and will embarrass both the contractor and the architect when the contractor has to ask the architect to correct his or her behaviour.

If a clerk of works or other site inspector is appointed, the architect should meet and discuss progress with the inspector in private before beginning inspections or attending to matters raised by the contractor. The architect and site inspector should then return to the site agent and agree how the architect is to proceed with the visit. Not to meet the inspector first would be discourteous to the inspector, undermine the inspector's authority, and leave the architect ignorant of valuable insights that the inspector may have.

During part of the visit the architect must inspect the work alone, or accompanied only by a site inspector. It will also be necessary to visit at least parts of the site with the site agent to discuss matters noted by the architect and to answer queries raised by the contractor. A workable procedure might be for the architect first to inspect the site alone, and then to go round with the site agent, when the architect may ask questions, point out matters requiring the contractor's attention, and discuss the contractor's own queries.

It is important that the architect does nothing on site that may compromise lines of authority or communication. When the architect is carrying out inspections alone he should politely greet everyone he meets but should avoid entering into discussions with tradesman or foremen about the work in progress. If site etiquette is to be strictly followed the architect should not discuss such matters directly with tradesman or foremen even when walking around the site with the site agent, but should address all remarks to the site agent. The site agent should then turn to the tradesman, confer with him as necessary, and turn back to the architect with a response. However, such a procedure can be unnecessarily formal, and it is likely to be appropriate for the architect to ask questions directly of operatives provided the site agent is present. Indeed, intelligent questions asked of tradesmen can lead to helpful and informative answers, add to the architect's knowledge of construction techniques generally, and improve the architect's relationship with the contractor's site team as a whole.

Under no circumstances should an architect address criticism of work or any other form of censure directly to an operative – or even a trade supervisor. If the architect sees evidence of bad practice or work that is not in accordance with the contract he or she should first make sure of his or her ground by looking very carefully at it him or herself, and then take up the matter as discreetly as possible with the site agent. It is for the site agent to decide who to involve in any discussion with the architect.

The architect should always make a point of complimenting good work or progress. If etiquette is to be strictly followed, praise, like criticism, should be directed to the site agent – but if possible should be done within earshot of the tradesmen responsible for the work. The architect should then praise the tradesmen directly.

If in exceptional circumstances work is discussed with an operative or subcontractor in the site agent's absence, the gist of the discussion must be reported to the site agent as soon as possible.

When walking around the site with a clerk of works or other inspector and the site agent, the architect should maintain the inspector's authority by – as appropriate – addressing queries to the inspector in preference to the site agent or operatives.

The architect should be sensitive to the relationships between the wide range of people under the control of the contractor. Such people will include main and subcontractor's office- and site-based managers, foremen, supervisors and operatives. All communications with people on site are best made in the presence of the contractor's person in charge.

Neither the architect nor site inspectors should mark with chalk or markers, or otherwise deface defective work. They probably have no right to do so, it is unnecessary, and it will needlessly annoy the contractor. The contractor may choose to do so himself if he wishes.

The relationship between the inspecting architect and the site agent is most important. It should be a relationship of mutual respect. The site agent should respect the authority of the architect as the representative of the architect's client; the architect should respect the authority of the site agent as the contractor's person in charge. A balance should be struck between strained formality and over-familiarity. The inspecting architect and site agent should meet on an equal footing, and should feel free to consult each other about queries and potential problems.

The architect should appear grateful to the contractor for raising genuine queries, and should thank the contractor for pointing out any errors in the architect's drawings or specification, or potential problems with the practicability of the architect's design. An alert and intelligent contractor who is made to feel appreciated can often save the architect from the consequences of potentially embarrassing oversights or errors.

Under no circumstances should the architect put him or herself in a position in which he or she could be accused of being too friendly with the contractor. The architect should not socialise with the contractor, either before or during the contract, except for occasional formal events such as a Christmas lunch. On such occasions the architect's client should also be invited.

An amicable relationship with the site agent or anyone else on site is of no value if it is maintained by the architect's being lenient at the expense of the client.

The architect will lose respect if he or she appears unfair, inconsistent, untrustworthy or dishonest. The architect should ensure that he or she never either puts himself or herself in a position where he or she may have to go back on his or her word, or makes promises that he or she may be unable to keep. The architect should not be seen to try and blame others for his or her own mistakes, or otherwise wriggle out of his or her responsibilities.

Under no circumstances should the architect complain about the client in front of the contractor.

6.4 Safety on site

As well as complying with basic health and safety etiquette it is absolutely essential that the architect pays strict attention to any additional rules laid down by the contractor in connection with health and safety.

The architect, however, must also take responsibility for his or her own personal safety. The architect should ensure, before climbing a ladder, that it is properly secured, and that before using scaffolding it is properly completed. When climbing ladders or walking on scaffolding the architect should always have both hands free. If the architect is in any doubt whatsoever about the safety of any means of access he or she should check with the contractor before using it. Under no circumstances should the architect use an unsafe means of access. If safe access to an element of work that the architect needs to inspect is not available the architect should insist that the contractor makes safe access available.

When moving about the site the architect should always look where he or she is going, even if moving only a few steps, paying attention both to where he or she is stepping and to what he or she is holding onto. It is extremely dangerous to walk around any part of a building site while looking at the building work, at drawings or at a checklist. The architect should never lean on scaffolding handrails or misuse any means of access or safety equipment.

6.5 Personal possessions

People working on building sites are no more dishonest than people working elsewhere. However, by their nature building sites can offer tempting opportunities to the casual thief, perhaps passing himself off as a delivery man or other legitimate visitor to site. Money and other items of value should therefore never be left unattended on site. Such items will need to be kept either in the pockets of clothes or in a backpack worn while walking around the site. The architect should never leave valuables in a coat or bag draped over a chair in the site office. Valuable possessions not related to the site visit should be left in the architect's office or at home.

6.6 What should the architect do if contractor's personnel behave in an uncooperative, obstructive or intimidating manner?

If the contractor has been carefully chosen it is highly likely that everyone with whom the architect comes into contact on site will be both cooperative and pleasant to work with. However, if the architect comes across uncooperative or obstructive conduct he or she should nevertheless continue to act in accordance with the guidelines on conduct outlined above. If the problem persists, such that the quality of the building work or the ability of the architect to do his or her job is affected, the architect should consider asking the contractor to remove from site those causing the offence.

It is highly unlikely that the architect will be subjected to intimidation of any sort, but it does happen on extremely rare occasions. Anything more than can be easily deflected by a judicious use of humour must not be tolerated. The architect must immediately take up such a matter with the contractor and if necessary insist that the offending person be removed from site. In extreme cases it may even be necessary to involve the police.

In most cases the people with whom the architect comes into contact on site are helpful, informative and extremely enjoyable people with whom to work, making inspecting work in progress a most interesting and satisfying part of the architect's job.

6.7 Building up knowledge

The architect can be most effective in dealing with the contractor on site if the architect knows what he or she is talking about. The architect does not need an encyclopaedic knowledge of all the building trades, but does need a general knowledge of basic construction techniques and trade skills. Before visiting site the architect should be familiar with the relevant parts of the contract documents, and must check up on matters for which he or she should look out, including – as appropriate – referring to literature from manufacturers or advisory organisations. However, matters will arise on site that require knowledge, understanding or experience that the architect cannot be expected to have. In such cases the architect should gain as much insight as possible by asking questions of the site agent and tradesmen. The architect should then consider checking the contractor's advice by referring to published guidance, building product manufacturers, trade organisations, colleagues and any other appropriate source of guidance.

6.8 How should the architect deal with queries raised by the contractor on site?

It is inevitable that during the course of the inspecting architect's visits to site the contractor will raise with the architect queries in connection with construction technicalities, apparent discrepancies within the contract documents, unforeseen site conditions that appear to make design details impracticable, and other matters. The answers to many of the contractor's questions will be obvious to the architect, who in such cases will be able to answer immediately.

However, there will be other questions to which answers may not be immediately clear. The architect must be extremely careful when answering such questions. The complexity of even the smallest of building projects is such that an apparently harmless change can have a costly effect on a following trade, infringe Building Regulations or other statutory controls, or conflict with the client's brief.

The architect cannot reasonably be expected to have in mind the reasoning behind every design decision made during the earlier stages of the job, but may nevertheless feel put under pressure to answer quickly – either by the contractor, or by the architect's own desire to appear competent and decisive.

In practice there are few questions in building that need immediate answers – whatever the contractor may say. To answer questions too quickly in order to appear in control could lead to the architect's authority being undermined rather than strengthened, to the project being delayed, and to additional costs to the architect's client. The professional way to deal with such a situation is to:

- consider the question carefully while still on site
- ask the contractor and trade operatives for their suggestions, and discuss with them any immediate points arising
- tell the contractor that it is necessary to give the question some consideration in the office, after which it will be promptly answered
- return to the office, ascertain all the facts, look at all the relevant contract information, and think through the question and its possible answers, taking into account all possible implications
- discuss and check conclusions with colleagues
- formulate an answer and pass it on to the contractor without delay.

6.9 Can the architect do too much inspection?

If the architect or a site inspector carries out thorough and frequent site inspections there is always a risk that the contractor will be tempted to neglect his own quality control activities, relying on the architect or site inspector to notice defects. It could therefore be argued that reducing the frequency of inspections by the architect or site staff would actually increase the vigilance of the contractor. The argument may be reasonable in theory. However, the architect in practice would be taking a high risk if in response to a contractor's failure to perform his quality control responsibilities the architect simply reduced the frequency of his or her own inspections. In any case the contractor should not only be picking up defects but should be preventing them from happening in the first place.

The best way of dealing with such a situation is for the architect to:

- carry out checks at random so that it is not possible for the contractor to predict when or where the architect will carry out inspections
- if defective work is found, instruct the contractor both to make it good, and to check and make good all similar defective work
- carry out further spot checks to ensure that the contractor really is making good as necessary, including opening up and testing as appropriate, until it can be reasonably assumed that no further defective work remains
- increase monitoring of the contractor's quality control activities and take action as necessary in the event of failures.

6.10 What should the architect do if the contractor tries to persuade the architect to accept defective work?

When defective work is found the contractor may attempt to persuade the architect to accept the work or to accept an unsatisfactory remedial proposal. The architect may out of sympathy for the contractor, or because of his or her own diffidence, feel tempted to accept the work as proposed by the contractor. The architect has no power to accept such work, and to do so without properly advising his or her client would be professional negligence. The architect must insist that defective work be made good. To do otherwise would:

- suggest to the contractor that lower standards will be acceptable generally
- undermine the architect's authority
- risk either a compromise in quality, or a building failure as a direct or knock-on effect of the defective work
- invite legal action from the architect's client.

6.11 What should the architect do if the contractor suggests replacing a specified material or work process with an alternative?

The architect must treat such suggestions with extreme caution. As advised above (see p. 71), changes to designs and specifications during construction can have serious consequences that are not easy to foresee. Furthermore, to minimise the risks involved it is likely that the architect will need to spend a significant amount of time, at his or her own expense, checking the contractor's proposals. It is therefore best, unless the contractor is proposing a very obviously like-for-like substitute for an item of minor importance, or the architect's original proposals are shown to be flawed, for the architect politely to reject the contractor's proposals.

6.12　What should the architect do if the contractor is slow in making good defective work?

The contractor may deliberately delay the making good of defective work in the hope that the architect will either get fed up with asking about it or will forget it, or that if it remains for long enough it will become too late to put it right. Occasionally the contractor may, with similar hopes, ignore the architect and continue to complete work that the architect has told him is wrong. In either case the architect must be adamant, must persistently bring the matter up during visits and meetings and make sure it is included in minutes, must ensure that payment for the work is not made, must issue to the contractor any relevant notices for which the contract provides, and must ultimately consider with the client arranging for another contractor to attend to the defective work, deducting the associated costs from the first contractor's final account.

6.13　What should the architect do if suspecting the contractor has covered up incomplete or defective work?

The architect should try to avoid such situations by timing inspections so that important work is seen before it is covered up. However, if the architect thinks that, despite his or her efforts, defective work has been covered up, the problem is simply resolved by instructing the contractor to open up as necessary to enable the architect to carry out a spot check. Adequate provision to allow for the cost of such opening up and any consequent effects on progress should have been made in the contract documents. If defective work is found, the architect should then use the relevant powers provided by the contract to ensure to his or her reasonable satisfaction that all defective work has been made good.

7 Inspecting the work

7.1 Inspecting generally

Before visiting site to carry out inspections the architect must prepare a list of the matters that he or she proposes to check. Such checklists must be compiled by referring to the contract production information. Published checklists may help, but it must be borne in mind that they can provide only:

- general guidance as to the kind of thing for which the architect should look out
- confirmation or clarification of the standards of workmanship that an architect can generally expect of a contractor
- a reminder of anything obvious that the architect may have forgotten.

The architect does not need to check everything but must form a considered opinion on what it is essential to check and what may be given lower priority. Such considerations should be made carefully: the architect may later have to justify his or her decisions in the case of a dispute.

Some matters can be checked by a quick glance. The inspection of others will require more effort. If the architect feels uncertain about how to go about checking something he or she should consider asking the contractor to demonstrate compliance. The contractor should be doing his own checks anyway, and should not object to doing them in the presence of the architect or, within reason, repeating them for the architect's benefit.

Regard should be had to special tolerances.

7.2 Preliminaries

Once on site, the architect should consider checking that:

- the building control officer and other officials have been notified as appropriate
- the contractor's site management team or person in charge is on site
- welfare facilities are set up, are of a high standard, and are being used
- the site is tidy, and materials are logically and accessibly stored
- there are sufficient operatives on site to make adequate progress
- site security, safety provisions and temporary protection are as specified
- a schedule of conditions has been prepared by or with the contractor.

7.3 Demolition and site stripping

The architect should ensure that the consultant structural engineer inspects demolition work as appropriate.

During demolition and site stripping the architect should be wary of potential claims by the contractor for additional payment (on the basis, for example, that because existing ground levels are higher than shown by tender information more excavation is required). The architect should check and record the evidence presented in support of such claims before it is removed or covered up.

If work involves a listed building the architect should also check that the contractor is not damaging or removing structure, features, finishes, fittings or building fabric generally to a greater extent than has been expressly permitted by the local authority.

The architect should consider checking that:

- existing services have been surveyed and tested as appropriate
- the work is being properly supervised
- sufficient opening up of existing buildings has been carried out to ascertain the scope of structural repairs, timber treatment, plaster replacement and other necessary remedial work
- the contractor clearly understands what is to be demolished, what is to be retained, and what is to be protected – and that adequate temporary protection is in place. (The architect should consider paying an extra visit or two to check.)
- topsoil is being removed to a depth of 150 mm – or lower if necessary – but if to be kept for reuse is not being dug up and contaminated with soil from other strata. (Topsoil for reuse should not be stored in piles higher than 1500 mm, and should be protected to avoid contamination.)
- trees are adequately protected – from impact damage to trunks and branches, and from root damage (by excavations or overloading of soil above).

7.4 Setting out

The architect should be satisfied that the setting out of the building on the site has been done correctly. The best procedure is for the architect to ask the contractor to demonstrate how the setting out has been done. The contractor should:

- show that the building is correctly related to site boundaries, building lines and other relevant features of the site

- lay tapes against profiles and lines so that the architect can see that dimensions are correct
- demonstrate with reference to diagonals that the building is square.

Throughout the construction period the architect should check that boundary conditions are not being infringed either horizontally, or vertically (by digging too deep or building too high).

7.5 Substructure, below-ground drainage, and structural concrete

7.5.1 Generally

The person best qualified to check structural work on behalf of the architect's client is the consultant structural engineer. The architect should therefore ensure that the engineer carries out inspections as appropriate. However, the architect should not simply turn a blind eye to structural matters but must carry out his or her own inspections (although the architect cannot be expected to do so with the same expert eye as the engineer).

The architect should check that thermometers by which temperatures are being monitored are kept out of direct sunlight.

7.5.2 Piling

Piling should be properly supervised by the contractor, using an experienced and properly qualified specialist supervisor. It should be inspected by the consultant structural engineer. The architect should ask the contractor to demonstrate that the piles are accurately positioned.

7.5.3 Excavations

The depths of excavations for foundations are designed to ensure that the foundations bear on stable ground and are below frost level. Best practice is for the contractor to dig the trench down to the last 50–150 mm, call the building control officer to inspect it, remove the last layer just before the building control officer arrives, and pour the concrete immediately after he leaves. Architect and engineer should time their inspections accordingly. Excavations should not be left open for more than a day or so.

The architect should consider checking that:

- adjoining structures are not being undermined. (If there is any risk of collapse the contractor should be instructed to stop work and take appropriate remedial action. The structural engineer should be consulted immediately.)
- any voids discovered on a boundary with adjoining land are dealt with under proper party wall agreements (and not tacitly filled with new concrete or other material so as to constitute a trespass)
- trench support appears to be properly carried out and adequate to ensure safety of operatives and adjoining structures – consulting the structural engineer as necessary
- excavations are protected so as to avoid deterioration of faces and risk of spoil falling from sides into concrete
- excavations are not surcharged by plant, spoil or materials being placed too close
- centres of foundation trenches align with centres of walls
- depths of foundations are correct. (If the contractor has dug too deep, or needs to dig too deep because the ground at the correct level has been allowed to become muddy or wet, the architect should consult the structural engineer before allowing the contractor to proceed.)
- bottoms of trenches are uniform, firm, level, and free from lumps, foreign matter, mud and water
- sides of trenches and any steps in the bottoms have straight and vertical faces. (Upper strata or dark streaks dipping below the bottom of the trench, or the presence of tree roots, should be discussed with the structural engineer before concrete is poured.)
- steel pegs should be securely fixed to the bottom of trenches and levelled to ensure foundations are of the correct thickness and level. (Wooden pegs should not be cast in, as they will rot.)
- special precautions as required by the structural engineer have been taken where heights of steps exceed the depth of concrete
- the contractor is not pumping water into drains that may become silted up.

7.5.4 Below-ground drainage

Below-ground drainage must be thoughtfully specified. Materials, bedding, surrounds, and details at building entries and chambers are likely to be of different types according to the distance of pipes from or below the surface or buildings. The architect should inspect with careful reference to the specification, and should witness tests – which should be carried out strictly in accordance with the specification. Whether or not the architect is present, he or she should ask the contractor to provide full records of every test, including details of section tested, date, type of test and personnel carrying out and witnessing the test.

Before pipework is tested the architect should consider checking that:

- above-ground drainage connection sockets are in the right place
- falls are as specified (generally 1:40 for a 100 mm pipe, 1:60 for a 150 mm pipe), and in the right direction
- bedding or trench bottoms evenly support pipes
- bedding is of specified depth, width and material, and does not contain large lumps of soil
- concrete beds are of specified depth and width
- where specified, compressible board is fitted at the upstream face of sockets, and extends through the full cross-section of the concrete bed and surround
- pipes are not deviating from line beyond specified tolerances
- open ends are sealed before backfilling
- flexible joints are fitted within 300 mm of chamber walls, and adaptable couplings rather than rigid joints are used at connections to existing pipework
- building entries are as specified including, as applicable, clearance around pipes, masking to prevent entry of vermin or fill, flexible joints located within 150 mm of wall faces on both sides, rocker pipes no longer than 600 mm
- rest bends are supported as specified.

After pipework is tested the architect should consider checking that:

- side fill is placed to extend the full width of the trench, and is compacted in layers of 100 mm; and at least 150 mm of compacted top cover is applied
- gullies and rest bends are backfilled with granular fill or encased in concrete as specified, and gullies are at the right level
- drains are kept free of rubbish until practical completion.

In connection with inspection chambers and manholes the architect should consider checking that:

- prefabricated chamber sizes and backfill are as specified
- concrete bases are at least 150 mm thick
- brickwork joints are fully filled, flush pointed and not greater than 6 mm; bricks are class B engineering bricks or otherwise as specified; and mortar is 1:¼:3 cement:lime:sand
- precast concrete chamber sections are bedded and surrounded as specified
- channels are of specified material and section to suit angle of entry, laid to specified falls, bedded and fully pointed in 1:3 cement:sand mortar, and fixed with branches entering chamber at half-pipe level of main channel

- benching rises vertically from main channel to height not lower than soffit of outlet pipe, then slopes at about 1 in 12 to walls, and is finished with steel-floated 1:3 cement:sand mortar within 3 hours of forming
- cover frames are bedded in 1:3 cement:sand mortar, and double-seal covers are fitted internally.

7.5.5 Hardcore

The purpose of hardcore is to provide a base that is of continuous and consistent firmness. It is therefore important that it be free from rubbish that can cause dry rot or that can deteriorate to leave voids, and that it comprises material of uniform size so that it can be properly compacted. Nevertheless, the contractor may be tempted to incorporate rubbish or use demolition waste materials to avoid spending money on removing them from site and importing new hardcore.

The architect should consider checking that hardcore is:

- not frozen and the ground not frozen when hardcore is laid
- of material of the specified size (normally to pass through a 75 mm sieve)
- completely free from timber or timber products, roots or other organic matter, plasterboard, metals, big lumps of concrete, other unsuitable rubbish, and cavities
- mechanically compacted between layers – normally of maximum thickness 150 mm below slabs and 225 mm at the bottom of foundation trenches – strictly in accordance with the specification, including the use of a roller or compactor of the correct type and weight
- of the correct overall thickness (usually between 100 mm and 600 mm)
- blinded as specified, with no protrusions.

7.5.6 Damp-proof membranes and insulation below slabs

The architect should consider checking that:

- blinding is compacted and levelled
- membranes are:
   - of the correct gauge of polythene or correct thickness of other material or coating
   - laid with joints lapped by at least 150 mm and sealed
   - not punctured
   - lapped and sealed at penetrations

- properly fitted into corners and at upstands, with enough slack material to prevent stretching when concrete is poured
- laid with enough additional material at edges to lap with wall damp-proof courses
- kept clean, and treated with respect by the contractor – not stamped on, kicked, or scuffed with heavy boots
• insulation is:
  - tightly butted at joints with no gaps
  - effectively supported at edges
  - cut tightly around penetrations
• insulation and membrane are not damaged while placing and finishing concrete.

7.5.7 Reinforcement

Poor workmanship in connection with concrete reinforcement is unfortunately common. The architect should consider checking that reinforcement is:

- adequately stored to avoid deformation before fixing
- not bent crudely, nor without the use of a proper machine, nor in temperatures below 5°C
- fixed accurately in the correct plane (at the bottom of a simply supported slab, but at the top of a cantilevered slab, for example) and correctly orientated with the right bars outermost so that the stronger bars have the maximum possible load-bearing effect
- of bars of the correct type of steel – by consulting with the structural engineer. (High bond stress, high yield steel has bobbles; low bond stress, high yield steel is twisted; mild steel is just a straight bar.)
- not fixed with bars so close together that aggregate cannot pass between them
- fixed with adequate or specified laps (for welded fabric typically 40 × main bar diameter for end laps, and 30 × transverse bar diameter for side laps; for bars at least 34 × bar diameter for tension bars and 27 × diameter for compression bars) properly secured with adequate ties, with wire or clips projecting inwards
- fitted with cover spacers or chairs of purpose-made steel, plastic or concrete not spaced too far apart (normally not more than either 50 × the bar diameter for slabs, or at 1000 mm centres for slabs and beams)
- generally located, supported, spaced and secured such that the reinforcement is in the right position and that cover to all faces is adequate (at least 75 mm where concrete is against the side of an excavation), and that displacement will not occur when concrete is poured or vibrated

- immediately before concrete is poured, free from loose rust or other scale, mud, mould-oil or grease that would prevent adhesion of the concrete.

7.5.8 Formwork

Wet concrete is very heavy indeed, and needs to retain water to cure. Shuttering must therefore be capable of supporting very heavy loads without deflection, and must be tightly constructed. Shuttering must be especially strong if the concrete is to be vibrated.

It is likely that shuttering for visual concrete will need to be specially designed and specified, and special care should be taken during its inspection.

The architect should consider checking that shuttering:

- boards are treated as specified before reuse, and are not reused too many times
- is set out correctly and is accurate within specified tolerances
- appears adequately strong with joints closely fitted or sealed (Soffits should be well propped but special attention is needed to sides – which if not very firmly supported will bulge and shear immediately concrete is poured.)
- inserts for cut-outs, holes or chases, and temporary joint fillers for movement joints are in place and properly secured
- is free of sawdust, chippings, nails, wire, mud, water and rubbish before concrete is poured
- is clean, and evenly and thinly coated with release agent of uniform or specified type immediately before concrete is poured.

7.5.9 Concreting

During the construction of concrete frames special attention should be paid to critical tolerances. Where the subsequent fixing of cladding or other elements depends upon the achievement of special tolerances the architect should ask the contractor before pouring concrete to demonstrate that the tolerances will be achieved. Tolerances should be checked after shuttering has been removed.

For most applications concrete is delivered to site ready-mixed, although small quantities may be mixed on site. The architect should ask the contractor to provide, for each delivery of ready-mixed concrete, a certificate giving actual weights of aggregate, cement and water used. (Delivery notes can be used to check slab thicknesses by dividing the volume of concrete delivered by the area of the slab.)

The architect should consider checking that:

- materials are stored correctly. (Bagged cement should be stored in a dry, frost-free shed or building – if too fresh, and therefore hot, use should be postponed, and if lumpy, or not used by the date on the bag it should be condemned. New deliveries should be placed behind old deliveries. Different aggregates should be physically separated, be stored on hard, clean, free-draining bases, and be covered adequately to give protection from frost and contamination.)
- construction and daywork joints are located strictly as agreed with the structural engineer, and vertical stop boards are provided
- overlap at steps is, for strip foundations, more than 300 mm and not less than the depth of the concrete, and for trench fill foundations more than 500 mm and not less than twice the depth of concrete, with proper shuttering used
- any timber pegs are removed before pouring foundations or slabs
- hardcore or blinding with which concrete is to come into direct contact is wetted immediately before concreting, or preferably covered with 1200 gauge polythene underlay, lapped 250 mm at edges, to prevent premature loss of water
- concrete is poured when conditions are neither too hot nor too cold. (If concrete is poured when it is too cold the chemical reaction by which the concrete cures will not take place, and the concrete when it solidifies will have no strength. Frost-damaged concrete will spall and disintegrate at edges and faces, and should be broken out immediately. If concrete is poured when it is too hot the water within the concrete will evaporate too quickly to allow the concrete fully to cure and thereby gain its full strength. Engineers' specifications typically permit concrete to be poured in cold conditions only when the temperature is 5°C and rising – which is likely to be late morning before temperatures start falling in the afternoon; and in hot conditions only when the temperature is below 30°C.)
- the mix is tested as advised by the structural engineer or as specified. (If concrete is being mixed on site the quality of aggregates and water, the processes by which the quantities of materials in the mix are measured, and the process of mixing should all be checked. If concrete contains too much water it will not be strong enough, but if it contains insufficient water it cannot be efficiently compacted around reinforcement and in corners. The amount of water will depend upon the shape of the aggregate and the function of the concrete: aggregates with sharp rough edges need more water than smooth

round aggregates; heavily reinforced work where compaction is difficult needs more water than mass concrete. If a spade stands up on its own in concrete for a ground-bearing slab the amount of water is probably about right. Slump tests should be carried out and test cubes taken as advised by the engineer or as specified. Constant water and cement content are critical to consistent appearance of visual concrete.)

- concrete is poured promptly – typically within half an hour of delivery or mixing on site. (The architect should inspect preparations the day before pouring to enable the carrying out of any necessary remedial works before concrete is ready.)
- the risk of segregation is minimised by transporting concrete across the site smoothly, and discharging it from the dumper, barrow, or chute directly into its placing position, without being dropped from a height – even into deep trenches or columns (where pumped concrete should be used)
- concrete is placed in one continuous operation between construction joints
- compacting with vibrators is carried out with the type specified
- deep pours are made in layers of specified depth, typically 300 mm, compacting between layers, ensuring that each layer merges with the layer below, but avoiding segregation, and taking special precautions to ensure that joints will not be visible where concrete is to be exposed fair faced
- concrete is adequately compacted around reinforcement, cast-in accessories, and into corners – but not so compacted that aggregate is separated from grout. (Compaction is usually adequate when air bubbles stop rising to the surface. Concrete seriously honeycombed because it has not been properly compacted must be condemned – its strength, and ability to protect reinforcement and resist sulphate attack, will be significantly affected.)
- reinforcement, damp-proof membranes and formwork are not displaced during pouring, and that spoil from the side of excavations does not fall into the concrete – any of which will require breaking out of the concrete
- boards are removed and aggregate exposed at construction joints within 24 hours of casting, and the surface is well wetted and grouted immediately before recommencing concreting
- concrete is adequately cured using methods – usually by covering with polythene sheeting – and for periods as specified. (Polythene should be laid as soon as practicable after concrete is placed and compacted, should be removed only to complete finishing operations, and should then be immediately replaced. Surfaces that will be exposed to frost, and floor and pavement wearing surfaces, are likely to require curing for at least 10 days; other surfaces are likely to require at least 5 days.)
- in cold weather, concrete is adequately insulated from frost using methods and for periods as specified by the structural engineer

- at no time hard concrete is cut away without the structural engineer's authority
- shuttering is not struck sooner than advised by the structural engineer. (It may need to be kept in place for as long as 28 days, with shuttering to the soffits of beams left in place for longer than to the sides.)
- tie holes and blow holes are filled and finished as specified
- newly laid concrete is protected from site traffic for at least four days, and longer in cold weather
- green concrete is protected from weather, dirt, physical damage, indentation, shock and extremes of temperature
- in cold weather, voids cast into concrete are protected from water to avoid frost damage
- sealants are applied as specified
- elements have been built within specified tolerances.

7.5.10 Basements

Basement waterproofing systems should be very carefully specified; and must be inspected equally carefully, with close reference to the specification. Generally, the architect must check that:

- surfaces are prepared as specified
- chases and service entries are properly prepared
- waterproofing systems are applied strictly as specified.

7.5.11 Precast concrete floors

While inspecting precast concrete floors the architect should consider checking that:

- units are transported and stored the right way up
- bearing surfaces are level and flat
- bearings are adequate (typically for beams at least 90 mm on walls and 55 mm on steel, and for blocks at least 10 mm on beams)
- wall cavities are kept clear of protruding beams, and fill between beams is solid
- blocks are not damaged or cracked
- fine concrete infill is applied between adjacent joists and planks as specified, and grouting is properly applied to blocks
- restraint straps are fitted as specified
- floor voids and ventilation below ground floors are as specified
- provision has been made for all necessary services penetrations.

7.5.12 Damp-proof membranes and insulation above concrete floors

The architect should consider checking that:

- the floor surface is flat and clear of mortar droppings and other debris
- membranes are:
  - of the correct gauge of polythene, or thickness of other material or coating
  - laid with joints lapped by at least 150 mm and sealed
  - not punctured
  - lapped and sealed at penetrations
  - properly fitted into corners and at upstands, with enough slack material to prevent stretching when concrete is poured
  - laid with enough additional material at edges to lap with wall damp-proof courses
  - kept clean, and treated with respect by the contractor – not stamped on, kicked, or scuffed with heavy boots
- battens are provided as specified to support flooring at doors or below points of concentrated load
- insulation is:
  - fixed with no gaps between insulation boards, and that joints have been treated as specified to prevent migration of screed
  - effectively supported at edges
  - cut tightly around penetrations and turned up at perimeters as specified
  - protected with suitable boards to prevent damage by barrows and boots when screed mix is being transported and placed.

7.5.13 Screeds

As preparation and minimum thicknesses vary according to function and type, screeds should be carefully specified. The architect should consider checking that:

- preparation is carried out strictly as specified including any surface treatment and the removal of all dirt and dust
- bay sizes are as specified
- movement joints are located as specified, and movement joints in the base are extended through the screed
- gas pipes are properly protected and services generally are run in conduit or accessible ducts secured to the base as specified, residual depth over services is at least the minimum specified, and reinforcement above services is central in the depth of the screed
- reinforcement is of the specified mesh, positioned centrally in the depth of the screed, with adequate laps (typically 150 mm). (Chicken wire should not be substituted for the specified reinforcement.).

- the correct sand is used. (Sand for screeds is normally coarse sharp sand, which is paler than the soft sand used for mortar, and is very gritty to touch.)
- materials are properly weigh-batched, allowing for any wetness of the sand, and mixing is properly carried out with a suitable forced-action mechanical mixer. (Free-falling drum mixers are not recommended.)
- the mixer is cleaned so that solid material is not allowed to form on blades or paddles
- the material is laid as dry as possible, but not so dry that compaction is difficult. (It should not be possible to squeeze water out of a handful of the mix, and under no circumstances whatsoever should any water be forming puddles on the surface of the finished screed. Screed is laid before the mix starts to stiffen – which can be within minutes when laying proprietary quick-drying screeds in hot weather.)
- depth is as specified – controlled using screeding battens or levels while laying
- joints are located as specified, with edges of bays at joint positions flat and vertical
- screeds are very thoroughly compacted over their entire area, in layers of equal thickness, with the surface of the lower layer roughened before immediately laying the upper layer
- screeds are cured as and for the period specified (typically by covering with polythene, well lapped, and held well down at the edges with boards to protect against draughts) and protected with hardboard or other suitable material to prevent damage
- screeds are finished to suit the following floor finish: flat as specified, level, and without crumbling, cracking or curling
- there are no areas of partially or fully bonded screed that sound hollow in comparison with other areas when tapped with a stick. (If such areas are found they should be taken up and relaid.)

7.6 Structural steelwork

7.6.1 Generally

As with other structural elements the architect should make certain that structural steelwork is properly inspected by the structural engineer. After steelwork is initially erected there remains for the contractor a significant amount of work in checking setting out and levels, plumbing, levelling, tightening of bolts, and testing. The architect should ensure, when certifying payment for steelwork, that enough money remains uncertified to account for the cost of all such work that remains to be done.

Special attention should be paid to protective coatings, and to critical tolerances.

The architect should consider checking:

- fabrication of special or critical components in the workshop
- fixing holes are pre-drilled before galvanising, and galvanising is as specified, particularly at joints and at tight corners
- shop- and site-applied protective coatings are strictly as specified, including preparation, use of the right materials, and dry film thicknesses. (The architect should arrange for tests to be carried out if in any doubt. Site priming should be applied immediately after preparation. Coating should not be carried out in adverse weather conditions. Damaged priming coats should be repaired before applying subsequent coats.)
- members are straight, untwisted, and have smooth ends as necessary to spread load evenly, and to receive protective coatings where applicable
- the appearance of welds to exposed steelwork is as specified
- friction grip bolt bearing surfaces and surfaces of steelwork to be encased in concrete are left unpainted
- steelwork is stored on site so as to avoid risk of distortion, or damage to coatings
- concrete bases are at the correct level
- steelwork is set out as drawings, with beams at the right level
- stanchions are correctly orientated
- members are of correct sizes
- bearing positions and lengths on padstones or other supports are as specified
- plumbing and levelling is properly carried out and checked by the erector before base plates are packed. (The architect should consider asking the contractor for a certificate to the effect that setting out of columns on gridlines, plumbing and levelling of columns, levels of beams and tolerances generally have been checked by the fabricator. If appropriate the architect should consider spot checking that columns are plumb himself or herself by using a bob. Where the subsequent fixing of cladding or other elements depends upon the achievement of special tolerances the architect should ask the contractor to demonstrate that the tolerances have been achieved. In all cases the architect should stand back and look at how columns align with other columns, or with other existing vertical references.)
- holding-down bolts are at right angles to base plates, of length to suit the level of the concrete base, with at least a full turn of thread visible above the nut, with the base plate packed off the base with steel packers, leaving a gap of about 50 mm between base plate and base, packed with dry 1:1 cement:sand after plumbing and checking levels
- other bolts are square in their holes, long enough, and properly tightened, and that any tests have been carried out and certificates received as specified by the structural engineer
- fire protection is applied strictly as specified.

7.7 Timber structure

7.7.1 Generally

The architect should ask the contractor for preservative treatment certificates both for new timber treated off site and for existing timber treated in situ, and should consider checking that:

- timber is free from large dead knots, significant wane or shakes, and is straight and square
- timber is kept dry
- all cut ends are preservative-treated
- in-situ treatment of existing timber has been carried out as specified, with the building watertight, and clear of rubble and materials to ensure that access to all parts to be treated is easily available, and with all infected timber immediately removed from site.

7.7.2 Suspended timber floors and ceilings

The architect should consider checking that:

- joist spacing is as specified
- double joists are provided where required and bolted as specified, with washers behind nuts and bolts
- trimming around openings and projections is strictly as specified, including sizes and joints
- joists bear at least 90 mm on masonry and 45 mm on steel, and sit squarely on masonry, a wall plate, or a hanger and do not protrude into external wall cavities
- wall plates are bedded dead level
- ground floor wall plates are bedded on a properly lapped damp-proof course, with joists cut back 25 mm from external wall faces, with floor voids and ventilation as specified, including free flow of air between partitions, sleeper walls and structural cross-walls, with ventilation routes not obstructed or blocked by debris
- built-in joists are treated and wrapped as specified, with the spaces between joists solidly filled with masonry
- joist hangers are supported directly by masonry units or other construction without being packed up or bedded in mortar, and are of suitable strength and size to suit joists; bottoms of joists are not excessively notched to fit hangers; joists are fitted to hangers with less than 5 mm between the end of each joist and the back of each hanger, with hangers securely fitted tight

against the supporting wall, with a suitable nail in every hole, and with the tops of all joists level
- end joists are positioned about 50 mm from masonry walls
- strutting or blocking is strictly as specified, with folding wedges between end joists and walls
- notching and drilling of joists and other members is strictly in accordance with specified limitations
- insulation is supported and secured as specified, packed around services at penetrations, and completed – including between joists and walls – with no gaps
- below-floor pipework is insulated
- restraint straps are fixed without being cranked or accidentally bent, at specified spacings, with fixings as specified (typically 50 mm no. 10 wood screws or 75 mm no. 8 round nails), notched into joists precisely to the thickness of the straps, connected to at least three joists, with noggings between joists, and folding wedges between end joists and walls
- noggings or other supports needed for the fixing of WCs etc are provided
- acoustic insulation is fixed strictly as specified
- boards are of the specified material and thickness
- tongued and grooved strip boards are fixed with heading joints centred on joists or battens with joints tightly butted, at least two board widths apart on any one joist, and are cramped before nailing, with two nails per board per joist using nails of two and a half times the board thickness
- panel boards are laid with all edges continuously supported, with edge treatment, fixings and fixing centres as specified
- boards are fixed flat and with expansion gaps at perimeter and at penetrations as specified
- holes around services penetrations are sealed
- boards are protected from the weather and subsequent trades.

7.7.3 Timber framing and studwork

The architect should consider checking that:

- timber framing is assembled and sheathing fixed using fixings as specified, with sheathing boards of the specified material tightly butted
- timber panels are stacked vertically, out of contact with the ground, supported to prevent distortion, and covered to ensure that they are kept dry
- during erection, panels are adequately supported
- sole plates are properly located, bedded, packed and securely held down with specified fixings. (Bedding should be not be more than 12 mm thick,

continuous and the full width of the plate. Packing should be of durable, non-compressible material.)
- panels are vertical
- head binders overlap vertical joints in panels
- immediately after erection, sheathing is protected from the weather with breather membrane as appropriate
- unauthorised alterations to panels, and notches to studs for services, are not carried out
- holes for services are located on the neutral axis, are not of diameter greater than a quarter of the width of the stud, and are located between 0.25 and 0.4 times the height of the stud
- all junctions between structural members are tightly butted, with no studs too short, and no gaps between wall plates and floor or ceiling structure
- timber lintels are supported on cripple studs, or by being splayed half-housed into studs, and multiple lintels are fixed together as specified
- supports are provided for radiators, and for any heavy fittings or other objects that need to be fixed to the wall after finishing
- cavity barriers are of specified size, are fixed securely with continuity at joints and intersections as specified, and are fitted with damp-proof courses as specified
- breather membranes are as specified including location, material, fixings, horizontal and vertical laps, with horizontal laps arranged so that water drains outwards
- insulation is tightly butted, continuous, and properly secured and prevented from slumping as specified
- vapour control layers are as specified including location, material, fixings, joint support, laps, continuity and sealing to penetrations.

7.7.4 Roofs

The architect should consider checking that:

- wall plates are centred on inner leaves, fully bedded in mortar, half-lapped 100 mm at joints, and dead level and parallel
- jointing in individual members is only where allowed, and strictly as specified
- trussed rafters or other prefabricated components are strictly as specified, free of unauthorised adaptations and damage, and stored to ensure that they remain undamaged
- rafters or trusses are parallel, spaced at specified centres, plumb, and free of bowing
- joints between members and wall plates, and between members, are strictly as specified

- diagonal bracing is installed and fixed to trusses strictly as specified
- binders at ridge and tie levels tightly abut gable and separating walls at both ends (by using overlapping lengths), and are not subjected to unauthorised alterations (for example to clear a flue)
- lateral restraint straps are fitted at 2 m maximum centres along walls parallel to rafters, fixed to at least three rafters, with noggings between rafters and folding wedges between end rafters and walls, with fixings as specified (typically 3.35 mm × 65 mm round wire nails or at least four 8 gauge × 50 mm countersunk head plated steel wood screws), with downturn tight against cavity face of inner leaf of wall
- holding-down straps are as specified, including spacing, lengths and fixings
- tank support structures, trimming and bracing around access hatch openings, and provisions for services penetrations are constructed strictly as specified
- timbers are not located within 200 mm of a flue or within 40 mm of the outer surface of a masonry chimney.

7.7.5 Stairs

The architect should consider checking that:

- the rise is correct when thicknesses of floor finishes are taken into account
- timbers are of specified thicknesses
- risers and treads are housed and glued together at top and bottom of each riser; are provided with three or more glued blocks under nosings, and three long steel countersunk screws at the internal angles of treads and risers; are glued and wedged into routed grooves in strings; and are provided with additional glued blocks below junctions of tread and string – including below winders
- nosings are as specified
- rough bearers are fixed at top and bottom of stairs as specified, and brackets are provided as specified
- newels, balusters and handrail are fixed as specified.

7.8 Masonry

7.8.1 Generally

With new facing work sample panels should be prepared as necessary to experiment with the effects of different joint profiles, bonding patterns, mortar colours and masonry units. Precise details of each of the sample mortar mixes should be recorded at the time of mixing. Sample panels should be left for a week or so before choosing, as the mortar will change colour as it dries out.

A control panel should be built, the quality and appearance of which the work on the building itself is to match. The panel should comprise at least 100 typical units, and should establish the quality of the units themselves, the degree to which chips or blemishes are acceptable, the characteristics of joints, and quality generally. The panel should be retained and protected until completion.

A similar approach should also be taken to re-pointing: in an unobtrusive area of existing masonry trials should be carried out and a control panel established.

The architect should instruct the contractor to take down and rebuild any work that does not comply with approved samples.

A mock-up should be erected and approved before decorative patterns or other intricate work are attempted.

Before work starts the architect should advise the contractor that perpend and any back joints are to be fully filled with mortar. Cavity work and damp-proofing details to chimneys need careful detailing and special attention from the architect on site.

Protection of ongoing and finished work against the weather and impact damage is as important as the bricklaying itself – even light rain on new brickwork can have a seriously disfiguring effect.

Defects are noticeable from long distances at ground level, so it is important that immediately before scaffolding is struck all new masonry, repointing and other repairs are inspected methodically and carefully, paying particular attention to colour and texture.

7.8.2 Materials and storage

The architect should consider checking that:

- masonry units are ordered to be of a consistent colour. (Bricks from a single firing batch should match each other in colour, but may not match bricks from different batches. Such matters should be discussed with the supplier when the bricks are specified, and – if necessary – special orders should be made. Even bricks within the same consignment will vary in colour to a certain extent owing to the nature of the raw materials and the firing process. It will therefore usually be necessary to open and work from at least three packs at once to avoid bands or patches of slightly different colours.)
- masonry units are ordered to arrive on site just before they are needed (to avoid unnecessary exposure to the weather or risk of impact damage)

- masonry units are properly unloaded as close as possible to where they will be used, and stored out of contact with the ground. (Under no circumstances should they be simply tipped from a lorry into a pile. To minimise shrinking of concrete blocks after laying, and to avoid efflorescence in brickwork, masonry units must be kept dry – even after packs have been opened and bricks are being used.)
- masonry units are as specified and not damaged or defective. (Bricks that can be scratched with a thumbnail, have arrises that can be easily broken off, break easily when struck against another brick, are cracked or pitted, show unburned nodules of clay or lime, show gravel or stone, or have areas of partial vitrification are likely to disintegrate partially or fully and should be condemned. Bricks should have flat and perpendicular faces and beds, should not be twisted, should be of the correct dimensions within specified tolerances, should be of uniform colour unless specified otherwise, should have a clear metallic ring when struck with another brick, and should reveal a uniformly burned texture when split in half. Absorption can be tested by thoroughly drying, weighing, soaking in water for 24 hours, then weighing again. The first delivery of bricks should be checked and examples of unacceptable bricks kept on site as control samples. Samples from each subsequent delivery should be checked.)
- cement and sand are obtained from the same original sources throughout the job (to maintain consistency of mortar colour)
- cement is kept in a weatherproof store with a dry floor
- sand is kept on a free-draining base clear of other aggregates
- pre-mixed lime:sand is protected against drying out or excessive wetting
- rolls of damp-proof course material are stored on end to prevent squashing and distortion, and are kept in warm conditions in winter
- sealants for movement and other joints are not out of date, and are protected from frost and excessive heat or humidity.

7.8.3 Workmanship generally

The architect should consider checking that:

- masonry units are not frosted or saturated, and work is not carried out in temperatures below 3°C
- the wall is set out correctly
- all openings are set out at foundation or ground level to ensure that they are accommodated by the bond in the best way possible as brickwork rises. (It may be necessary to select bricks of special dimensional consistency for narrow piers or columns.)
- the mortar mix is as specified, and properly mixed consistently throughout the job. (A failure to measure proportions carefully and consistently can seriously

affect the durability and appearance of the finished brickwork. The correct sand should be used – it should leave a stain if rubbed between fingers. Sulphate-resisting cement should be used where specified. Sand and cement should be measured by volume, using clean and accurate gauge boxes. Allowance for bulking of damp sand should be made. Site mixing should take place in a proper mixer. The mixture should be placed in the mixer and clean water added while the mix is turning over. The mortar should be mixed for the correct amount of time – usually 3–5 minutes. Mortar should never be mixed directly on paving or the ground. If ready-mixed mortar or ingredients are used the architect should ask the contractor for copies of delivery notes. On larger jobs cube tests of mortar should be taken at regular intervals. At the end of the day all bankers, barrows, mixers and batch boxes must be hosed completely clean. Mortar left over must not be reused the following day.)
- additives are used only strictly as specified, and are properly measured and gauged with the water before the mortar is mixed. (Mortar with air-entraining additives should not be mixed for longer than recommended by the manufacturer – normally for no more than 5 minutes. Under no circumstances should additives be squirted in neat as or after the mortar is mixed, as the additives will not be evenly distributed within the mortar and the strength of the brickwork will be affected.)
- bricks are kept completely dry when stacked out on boards at ground level or on scaffolding
- facing work is kept free of mortar and clean
- perpend and back joints as well as bed joints are properly filled by fully buttering head faces when laying, and bricks are laid with frogs or deeper frogs upwards
- the bond is as specified
- the joint profile is as specified – consistently throughout the job. (To avoid a patchy appearance bricklayers working on the same site should all be using the same joint profiling technique.)
- damp-proof courses are of the specified type and fixed strictly as specified, including overlap with damp-proof membrane, mortar bedding, laps, detail at face of wall, and sealing at laps where resistance to downward movement of water is required
- perpends are aligned vertically (or at every fifth perpend with even variation in between so as to avoid a wavy effect)
- bed joints are horizontal, and gauge rods are used to ensure that courses rise consistently throughout the height of the wall
- the brickwork is vertical generally, and at external angles and reveals (using as appropriate a long spirit level, a plumb bob, or by simply standing back from the work and looking)
- brickwork is racked back rather than toothed when raising corners, and at temporary terminations

- brickwork is not built in single lifts greater than 1500 mm
- joints between new walls, new and existing walls, and walls and partitions are carried out as specified
- an adequate gap to allow differential movement is left below oversailing timber-frame-supported structures at eaves and verges, and below window sills of timber-framed buildings
- openings are plumb, square and of the correct size
- reinforcement is provided as specified
- lintels are fixed the right way up, level, properly bedded in mortar, with specified bearing length, on bearings of adequate integrity as specified
- separating walls are extended into roof spaces, with joints between wall and roof filled as specified
- slip ties are fixed at movement joints as specified
- compression joints and head restraints are formed as specified at heads of non-load-bearing partitions and brickwork cladding panels
- cladding panels and other brickwork panels are tied to structural frames as specified
- movement and fillet joint widths and depths are as specified, with sides flush and parallel. (Surfaces should be free of frost, dust, oil, grease, water and dirt before priming, and adjacent non-porous surfaces should be masked with tape – but non-porous surfaces should not. Surfaces should be fully primed, without excess, and backing of uncompressed width between 125% and 150% of joint width, and thickness of greater than 50% of joint width, should be fitted untwisted at even depth. Sealant should be mixed by the correct mixer at the correct speed and applied in temperatures between 5°C and 40°C, be correctly spread, and be separately tooled to produce a smooth, even, flat or slightly concave surface without sealant on adjacent surfaces. Triangular fillets should maintain contact of 6 mm to non-porous surfaces and 10 mm to porous surfaces, with backing where the gap to be sealed is more than 5 mm. Any masking tape should be removed immediately tooling is completed.)
- raking out for re-pointing is at least 20 mm (checking more than once during the course of the work).

7.8.4 Cavity work

The architect should consider checking that:

- the overall wall thickness and cavity width are correct
- cavity fill is provided from between 150 mm and 225 mm below the ground-level damp-proof course
- cavity trays are of the specified type, and are bedded in mortar, fully supported and sealed at laps, extended beyond intrusions, sealed to stop

ends, overlapped with damp-proof membranes, detailed at face of wall, and otherwise fixed strictly as specified. (Cavity trays must not be omitted over minor intrusions such as meter boxes, or air brick or other ducts.)
- damp-proof courses do not project into cavities (except vertical damp-proof courses at openings)
- weep holes are provided as specified
- wall ties are of the type and at the spacing specified, including additional ties at openings, movement joints and gables; are bedded as the units are laid at least 50 mm into each leaf and 25 mm from either face; are fixed to the studs rather than just the sheathing of timber-framed walls; and are the right way up, sloping downwards towards the outer leaf (to maintain slope adequately following shrinkage of frame in timber-framed buildings), with the drips centred on the residual cavity, and with the specified type of insulation retaining clip used
- lifting battens are used in cavities. (The battens should be raised and mortar carefully cleared off before fixing the next run of wall ties. Any mortar finding its way onto wall ties below the batten should be knocked off with a stick. At intervals of no more than 2 m bricks – or preferably internal blocks – in the course immediately above the base of the cavity, and above openings, should be bedded in sand to allow access for removal of mortar that has fallen or been knocked off. The mortar should be cleared before it has hardened – at least once a day. Boards should be used to protect full-fill cavities.)
- insulated cavity walls are built by building the inner leaf up to one course above the height of the next run of insulation boards, cleaning mortar snots from the cavity face, lifting and clearing the batten or board, fitting insulation boards tightly together between ties and securing with clips, building the outer leaf to the top of the insulation, and repeating until the wall is complete
- insulation is of the type specified; is squarely and neatly cut; is installed with horizontal joints coursed with wall ties, with vertical joints staggered, and with each board held firmly by four clips; is neatly cut around openings; and is tightly butted at corners
- built-in insulated cavity closers are built in with vertical damp-proof course projecting 25 mm into the cavity – and not pushed in after the wall is completed
- cavities below door thresholds are protected to prevent them from becoming filled with rubble and rubbish.

7.8.5 Protection of new brickwork

The architect should consider checking that:

- scaffold boards next to the wall are turned back at night or in wet weather to stop mud or spilled mortar being splashed onto the finished work

- vulnerable arrises to reveals, sills, arches, corbels, steps and plinths are protected from impact damage by cladding with clean softwood, plywood or hardboard – and by notices warning drivers of vulnerable overhead features. (The boards should be fixed where possible by tying wire around them, but if necessary they can be fixed to the brickwork with nails driven into mortar joints – which can easily be made good after protection is removed.)
- projecting details such as plinths and corbels are protected from mortar dropping from above. (Protection can be provided by building the edge of a polythene sheet about 20 mm into the mortar bed above the projection, with the sheet draped down over the projection. When the subsequent work is complete the sheet should be neatly cut off and the mortar made good as necessary.)
- immediately it is about to rain – and overnight – tops of walls and recently completed work generally are covered with heavy polythene sheeting. (An air gap must be formed between the sheeting and the brickwork by loosely laying battens between sheeting and brickwork. The sheeting must be long enough to cover all new brickwork below, and must be adequately weighted with boards or bricks to ensure that it is not displaced by wind. Any water running off the protection must be directed away from brickwork below.)
- in wintry conditions, recently completed work is covered to protect it from frost using layers of dry hessian sacking covered by heavy polythene sheeting – or preferably purpose-made insulated waterproofing material. (The protection should be adequately weighted down to ensure that it is not displaced by wind, and should remain in place for up to seven days until the mortar is cured.)
- in hot weather, recently completed work is prevented from drying out too quickly by covering it with layers of damp hessian sacking covered by heavy polythene sheeting. (The hessian should be sprayed with water as necessary to keep it damp, but should not be over-wetted, which would result in staining.).

7.9 Roof finishes

7.9.1 Generally

The functions of a roof are usually to keep water out and to keep heat in – and to do both as efficiently as possible throughout the design lifetime of the roof. There are many ways by which the functional requirements of a roof may be met, some of which will be long established and well understood by the construction industry, but others of which may appear unorthodox and will require explanation. The architect should ensure that the contractor fully understands both the overall design principles and the construction details – and that the message gets through to the roofer. Roofs must be constructed strictly as designed, and the architect should carry out careful and frequent inspections as the various elements of the overall roof construction are fixed. In particular, the architect should check that:

- precautions to prevent condensation are understood and fully implemented
- breather membranes, insulation, vapour control layers, and air leakage barriers are of the specified material, and are located and fixed as specified
- provisions for ventilation are constructed strictly as specified, and not subsequently blocked (by insulation, for example)
- junctions with fire-separating walls are as specified, particularly in boxed eaves and above wall plates (both of which must be inspected before closing up), and where walls abut the underside of roofing
- cover flashings are of materials, girth and length, are treated, are wedged and pointed, and clipped or otherwise fixed, and are joined – all as specified. (Cover flashings are essential to roof performance, and their execution can be less than perfect.)
- rainwater outlets are fitted with guards
- water in contact with copper is not discharged onto aluminium, zinc or galvanised steel
- lead is not in contact with aluminium in a marine environment
- on completion, the upper surfaces of roofs are cleared of all fixings, metal objects and other debris that may damage roofing, and adequate temporary protection is provided to ensure that damage is not caused by subsequent trades fixing aerials, gaining access to rooftop plant or carrying out any other potentially damaging activities.

7.9.2 Tiling and slating

Slates, in particular, are brittle and under no circumstances should be clambered over once fixed. It is therefore best if all fixing of aerials and other work above slating is, as far as is practicable, done before fixing slates. Access for fixing flashings must then be effected using proper roof ladders. The architect should consider checking that the following are as specified:

- tile and slate materials and sizes. (Upon delivery the architect should compare slates and tiles with samples, look at packaging, ask for delivery notes, and check for chipping at corners, cracking and other damage. Slates should ring if tapped by a metal object – a dead slate sounds dull – and should not show signs of water creeping up the slate if half submerged overnight in a bucket of water. Tiles, if broken in half, should show no striations, and should appear evenly burned without a light core and dark edges.)
- slates are sorted into three or four groups of equal thickness, and during sorting holed from underneath 20–25 mm from edges to produce a small countersink on the face without spalling – using a suitable machine
- slates are fixed in broken bond courses with thicker slates in lower courses, with all slates in each course of equal thickness, with the thicker end of

tapered slates laid at the tail, with tails aligned, with two nails per slate, with nail heads flush with slate faces, and with no more than 5 mm between slates
- the height of the tilting fillet and any fascia at the eaves is neither too low nor too high – such that while the bottom edge of the first full course will sit on the bottom edge of the eaves course without a gap, the upper face of the tilting fillet will nevertheless fall outwards towards the gutter
- counter-batten size and fixing
- tension of underlay
- underlay lap support, arrangement, size (typically 100 mm for sidelaps and for headlaps to roofs pitched over 35°, 150 mm for headlaps to roofs under 35° and at hips, and 300 mm past centre-lines of valleys) and sealing are as specified
- underlay is turned up by at least 50 mm and sealed at penetrations, around openings, and at perimeters
- underlay at eaves is fully supported by a tilting fillet and overlaps any fascia (typically by 50 mm) sufficiently to shed water into the gutter completely clear of the tilting fillet and fascia
- underlay detail at ridges, valleys and hips
- batten size, gauge, fixing, support at ends and location of joints are as specified
- levels of valley boards, and counter-batten and batten details at valleys and above openings, are such that if any water finds its way onto the underlay it is able to flow freely to the eaves, where it can discharge into a gutter
- provisions for ventilation are as specified
- lengths, laps, fixings and top saddles to lead valley linings are as specified
- details at abutments, and around openings, chimneys and penetrations are as specified
- tile and slate fixings generally are as specified
- at ridges, hips, valleys and verges tile and slate sizes, fixing, bedding and pointing are as specified. (Whole or one and a half size tiles and slates should be used. Slates should not be cut less than 150 mm wide.)
- distances between tiles and slates across the widths of valleys are as specified
- end ridge tiles and hip tiles are mechanically fixed, and bedded and pointed
- roof ladders are used to gain access after slating.

7.9.3 Profiled metal or fibre reinforced cement roofing

The architect should check that:

- sheets are checked for damage on delivery, and are carefully stored and handled so as to ensure they cannot be damaged in any way. (Potential hazards range from grit between coated metal sheets scratching the finish, to impact damage.)

- supports are of materials, are treated, are located, and are fixed strictly as specified
- claddings are fixed with laps or joints, fixings, washers, seals, and junctions with other parts of the external envelope strictly as specified – with no fixings, washers or caps missing
- provisions for preventing wind uplift are constructed strictly as specified
- eaves and ridge closure pieces are tightly fitted.

7.9.4 Fully supported metal roofing

The architect should consider checking that:

- all materials, including felt and wood rolls, are stored dry and under cover. (Lead rolls should be laid flat, and coils of other metals stood on end on flat bases clear of the ground. Pre-formed profiles should be laid flat and nesting, or on racks, and be protected from damage.)
- base construction generally, including gutters, is as specified
- falls are as specified. (The architect should check before and after laying of roofing.)
- before laying roofing the base is thoroughly cleaned to remove any projections – and is kept clean throughout the work
- underlay material and fixing are as specified
- at abutments parallel to roof structure differential movement between roof and walls is accommodated as specified
- type and thickness of roofing and clips are as specified – checking by comparison with a sample, or otherwise
- fixing of roofing is as specified
- lengths and widths of bays and gutter linings are as specified
- rolls, seams, drips, welts and joints generally are all properly formed as specified, without cracking of metal
- eaves, abutment, verge, ridge, gutter, penetration and expansion joint details are as specified
- roofing to receive any traffic whatsoever after fixing is completely covered and protected from damage.

7.9.5 Built-up and single-ply roofing

The architect should consider checking that:

- rolls of felt are stored upright on a clean solid surface
- falls, and base construction generally, are as specified

- base preparation, including the removal of all projections (to ensure that felts or membranes are not punctured), is as specified
- base and subsequent layers are dry (to avoid trapping moisture)
- at abutments parallel to roof structure, differential movement between roof and walls is accommodated as specified
- priming is as specified
- single-ply membrane and built-up layer materials and fixing are as specified
- laps (including size, and formation such that water is not directed by falls into joint) are as specified
- eaves, abutment, verge, ridge, gutter, penetration, expansion joint and rainwater outlet details are as specified
- surface protection is as specified
- roofing to receive any traffic whatsoever after fixing is completely covered and protected from damage.

7.9.6 Asphalting

The architect should consider carrying out unannounced spot checks as well as regular periodic visits during the laying of asphalt. The architect should consider checking that:

- falls and base construction generally are as specified
- a separating layer is provided as specified
- at abutments parallel to roof structure, differential movement between roof and walls is accommodated by independent upstands
- asphalt material is as specified – by reading label on blocks
- temperature of asphalt in mixers or cauldrons at no time exceeds 230°C. (Asphalt heated to higher temperatures will fail.)
- the thickness and number of coats to main surfaces are strictly as specified – by noting the depth of laying battens, by looking at bay edges, and by asking for a sample to be cut out and measured if in any doubt whatsoever. (Asphalt must always be laid in at least two coats to ensure that performance is not affected by the slight imperfections that would inevitably be present in a single coat.)
- joints in subsequent coats are staggered by at least 150 mm
- eaves, abutment, verge, gutter, penetration and expansion joint details are as specified – including primer or metal lathing as applicable; skirting support if asphalt is laid on insulation; thickness, number and application of coats to upstands; two coat fillets with face dimension at least 40 mm at 45° at upstand bases; surface protection; termination of upstands (ensuring chases are adequately formed); cover flashings; and rain water outlets
- sand rubbing is applied to the final coat. (If the asphalt looks shiny, sand rubbing hasn't been applied.)

- surface protection is as specified
- roofing to receive any traffic whatsoever after fixing is completely covered and protected from damage.

7.10 External wall finishes

7.10.1 Render

Before rendering is started the architect should ask the contractor to prepare a number of sample panels to experiment with the effects of different finish textures, and to establish a control panel. The architect should consider checking that:

- background construction, preparation and treatment are carried out as specified. (Backgrounds should be sound, stable, free from contamination, unsaturated, and keyed or primed as necessary.)
- corrosion-resistant lathing – preferably stainless steel – is fixed over junctions of dissimilar backgrounds
- beads are corrosion resistant – preferably stainless steel – and fixed with corrosion-resistant fixings at 600 mm maximum centres
- lathing is as specified. (It should be fixed taut, starting from the centre and working outwards using specified fixings, at least 5 mm off the background or support, with strands sloping downwards away from finished face of render. Supported laps should be at least 50 mm, wired together with a single row of corrosion-resistant wires at 150 mm centres, tied tightly with ends bent away from the finished face. Unsupported laps should be at least 100 mm, wired with a double row of wires at 100 mm centres. Laps at angles should be provided.)
- adequate measures are taken to protect against the effects of cold weather. (Internal work should be protected from air temperatures below 3°C. External work should not be carried out in air temperatures below 5°C and falling, or below 3°C and rising.)
- rendering is not carried out in full sun or hot, drying winds
- materials are accurately proportioned – normally in gauge boxes by volume, taking into account the moisture content of sand – and properly mixed in a clean mixer (except for very small quantities, which may be mixed on clean boards)
- coats are of specified thickness, are allowed as specified to cure properly between coats, and are keyed before applying subsequent coats
- render is carried into all reachable spaces, and is carried up tight to the undersides of window sills, and to frames at reveals and soffits
- details above openings in cavity walls and at horizontal damp-proof course level are as specified

- movement joints extend through the whole render thickness
- finish is accurate as specified (typically not deviating more than 3 mm under a 1.8 m straight edge) and is of consistent appearance to match the control sample
- finished render is cured as specified. (Portland cement gauged render is usually cured by covering with polythene sheet and spraying with water, or just spraying – during a minimum period of 3–4 days.)

7.10.2 Tiles

The architect should check that:

- background construction, preparation, and treatment are carried out as specified. (Backgrounds should be sound, stable, free from contamination, dry, and keyed or primed as necessary.)
- corrosion-resistant lathing – preferably stainless steel – is fixed over junctions of dissimilar backgrounds
- adhesives and grouts are of materials and are applied strictly as specified, including protection from the weather before and after grouting
- setting out is as specified
- movement joints are incorporated where specified, are of the specified width, extend throughout the tiling and background, are continuous throughout the tiling, are free of mortar or adhesive, and are sealed as specified
- joints are straight and level, free edges are level or plumb as appropriate, and tiles are laid flat, and flush at joints
- joints are fully and evenly grouted, and are slightly recessed
- edges are as specified
- tiles are adequately protected after fixing.

7.10.3 Claddings generally

Whatever the wall cladding, the architect should ensure that the contractor fully understands both the overall design principles and the construction details. In particular the architect should check that:

- panels are checked for damage on delivery, and are carefully stored and handled so as to ensure they cannot be damaged in any way. (Potential hazards range from grit between coated metal sheets scratching the finish, to impact damage.)
- precautions to prevent condensation are understood and fully implemented

- breather membranes, insulation, vapour control layers, and air leakage barriers are of the specified material, and are located and fixed as specified
- provisions for ventilation and drainage are constructed strictly as specified and are not subsequently blocked (by insulation, for example)
- provisions to accommodate differential and other types of movement are constructed strictly as specified
- supports are of materials, are treated, are located, and are fixed strictly as specified
- claddings are fixed with laps or joints, fixings, washers, seals, and junctions with other parts of the external envelope strictly as specified.

7.11 Windows and doors

7.11.1 Components generally

The architect should ask the contractor for a preservative treatment certificate for joinery, test certificates for fire doors, and should consider checking that:

- joinery to be painted is shop-primed, and joinery to be clear finished is shop-primed to bedding faces, with all other faces sealed
- components are not delivered to site before they are needed
- timber is of the right moisture content (checking with a moisture meter upon delivery, and immediately after fixing)
- items are stored in a weatherproof enclosure with a dry floor
- drips are formed in sills and thresholds
- some heat is applied to the building before fixing door leaves and other second fix joinery items – but not so much that excessive drying, shrinkage and cracking will occur
- all screw holes are fitted with screws of the correct type, driven in square.

7.11.2 Frames and leaves

The architect should consider checking that:

- timber frames are shop-primed as specified
- doors and windows are stored off the ground, under cover and protected from damage. (Doors should be stacked horizontally, adequately supported. Windows should be stacked either horizontally, or vertically with spacers.)
- damp-proof courses are fitted to sills
- any cut faces to timber frames are re-primed

- vertical damp-proof courses and sealant backing tapes are fitted to built-in frames before installing, and masonry is built lightly to abut installed frames
- frames to prepared openings cover damp-proof courses
- frames are plumb, square and correctly set out
- fixing blocks and spacers are used to prepared openings as specified
- frames are securely fixed. (Frames should be side-fixed at maximum centres of 450 mm for windows and 600 mm for doors, with top and bottom fixings within 150 mm of head and sill or threshold respectively. Where practicable frames should also be head-fixed.)
- frames are sealed all round externally, and any internal gaps are filled with expanding foam or similar as necessary to ensure that the installation is airtight
- fixed window frames are not used for access, or as support for scaffolding, boards etc
- gaps between window sashes and frames, between door leaves and frames, and between meeting stiles are not excessive
- draught stripping to external windows and doors, and roof access hatches, is as specified
- internal door frames are not fitted until immediately before plastering (to minimise the risk of damage from site traffic)
- stops to fire door frames are screw-fixed, seals as specified are fitted to heads and jambs, and fire-resisting door leaves are provided as specified
- ironmongery is fixed with screws of the right type, length and number, driven home without burring the edges of the slot
- sinkings for hinges, locks and keeps are neat and tightly fitting, with outer faces of metal and surrounding timber flush. (Packing should not be accepted. Back plates should be centred as specified, and be parallel with the frame of the door.)
- door furniture is set out precisely as specified
- window locks are properly adjusted, and casement stays correctly fitted
- all ironmongery is correctly adjusted and lubricated
- doors, windows, frames, glazing, architraves and ironmongery are protected adequately until practical completion.

7.11.3 Glazing

The architect should consider checking that:

- glass and glazed units are as specified – by looking at markings on glass, and by asking for documentary evidence such as delivery notes or guarantees
- materials and glass are stored in a dry, sheltered place out of direct sunlight, and are protected from damage

- rebates are primed as specified
- glazing compounds and their application are strictly as specified, including quantities and thicknesses
- setting blocks, location blocks and distance pieces are of specified material, and are used strictly as specified, wedging glass tight into sash as appropriate
- at least 3 mm of non-setting compound or capping sealant is visible between glass and frame, and beads are bedded strictly as specified
- timber glazing beads are fixed with fixings and at spacings as specified, without splitting beads
- safety glass, fire-resisting glass or other special glasses are fixed in specified locations.

7.12 Services

7.12.1 Generally

Very early in the contract the architect should check that all existing utilities have been surveyed, that design capacities required for each service have been established, that all necessary orders for new or replacement services have been placed, that all associated work has been properly incorporated into the contractor's programme, and that accommodation proposed within the building for the entry of incoming services is adequate and practicable. The architect should also ensure that as soon as services installation subcontractors are appointed they check and confirm that proposed services routes provide adequate space for all services and are otherwise practicable.

Materials, including ductwork, should be protected from the weather and risk of physical damage.

Where specialist consultant engineers have been appointed the architect should ensure that they carry out proper inspections. The architect or specialist consultant should witness tests of drainage systems and ask the contractor for a certificate. The contractor must provide test certificates for electrical installations.

Special attention must be paid to the passing of services through walls, floors and any other elements of fire-resisting construction. Holes in such structure should be neatly formed and not oversized. Intumescent collars and other fire-stopping materials should be carefully and properly installed. Dampers should be present in ductwork.

Practical completion should not be certified until services manuals have been completed as specified.

7.12.2 Electrical services

The architect should consider checking that:

- fittings are not installed back to back in separating walls
- vertical chases do not exceed one third of wall or leaf thickness
- horizontal chases do not exceed one sixth of wall or leaf thickness
- cables are fixed flat, without twists, and with plenty of slack at bends to allow for movement of structure.
- cables are run in centres of joists and never in notches
- cables in floor and roof voids are fixed to sides of structural members, at least 50 mm from top or bottom, with clips at 1 m maximum centres, supported on battens where they cross voids. (Cables should never be fixed to the tops of roof space joists.)
- cables run vertically are fixed to studs at 500 mm maximum centres
- cables in plaster are protected as specified
- lighting and power cables are segregated from fire alarm, emergency lighting, telephone, TV, hi-fi, door bell, door answering system and other cables
- electrical cables generally are segregated from hot water pipes and all other services that produce heat, smoke or fumes, and from services that cause condensation
- cables are not run in lift shafts
- cables are not run in insulation (which causes cable to overheat) or in contact with polystyrene (which degrades PVC cable sheathing)
- the setting out of all fittings is as specified – checking before plastering
- mounting boxes are square, level, at the right height, properly fixed, and fitted with grommets
- fire stopping is installed strictly as specified
- earthing, and main and supplementary equipotential bonding, are all carried out strictly as specified
- consumer units and distribution boards are properly and neatly labelled
- lightning protection systems are carried out as specified
- testing is completed as specified.

7.12.3 External above-ground drainage

The architect should consider checking that:

- temporary arrangements are made as necessary to ensure that new or existing masonry walls do not become saturated before new downpipes are fitted

- pipes and gutters are stored horizontally clear of the ground, with large quantities stored in racks
- materials are strictly as specified. (UPVC can withstand boiling water but PVC cannot, so can only be used for rainwater drainage.)
- gutters are fixed with corrosion-resistant screws to brackets at specified maximum centres (typically 1000 mm) and to each side of outlets, are set to specified falls (typically 1:350), are jointed as specified, and have guards as specified fixed to outlets
- roofing underlay is dressed into gutters
- downpipes are fixed with corrosion-resistant screws and plugs to brackets at the head (to support the swan neck or hopper), just below collars (to prevent pipe sections slipping), at the foot (to support the shoe), and otherwise at specified minimum centres (typically about 2 m), and are jointed as specified
- soil vent pipes are located at specified distances from windows and rooflights, and are terminated as specified
- horizontal waste pipes are adequately supported
- access eyes are fitted where specified.

7.12.4 Internal above-ground drainage, and heating and hot and cold water services

The architect should see that specified pipework tests are carried out, and should carry out his or her own checks on segregation, support, lagging and accessibility before pipework is covered up by floorboards and plasterboard. The architect should check for damage or incorrect ordering by the contractor of radiators, sanitary ware, taps, shower fittings etc as soon as they are delivered so that delays are minimised if replacements are needed. The architect should ask the contractor for records of tests and commissioning.

The architect should consider checking that:

- the water supply pipe outside the building is insulated and at least 750 mm below ground level
- the supply main below the ground floor of the building and within 750 mm of the external wall is insulated
- gas pipes embedded in concrete are protected by wrapping or coating in bitumen, and are run in a purpose-made duct, or a sleeve 5 mm thick, or foam pipe lagging or other resilient covering material
- gas and water supply pipes embedded in screed are protected by wrapping, or coating in bitumen
- pipes are sleeved as specified where passing through walls

- voids containing gas pipes are ventilated as specified and pipe locations marked where practicable
- balanced flue pipes are located as specified in relation to openings, gutters or painted surfaces
- flue pipes, where passing through walls, floors or roofs, are sleeved with non-combustible material maintaining an air gap between flue pipe and sleeve of at least 25 mm as specified
- conventional flue pipes are not within 25 mm of combustible material, are installed with sockets uppermost, are supported under every socket, are otherwise supported at intervals no greater than 1.8 m, have joints properly sealed as specified, have no sections shallower than 45°, and are terminated as specified in relation to openings, eaves, dormers, parapets and flat roofs
- copper tubes are properly bent using a machine
- joists are notched only strictly as specified – typically only within 250 mm of ends – and are fitted with felt pads to prevent noise when pipes expand, and steel protection saddles
- pipes are fitted with room for pipework expansion. (For example, bends should not be formed tight against joists.)
- capillary joints are properly formed, with pipes cut straight and properly reamed to remove burrs, and without too much solder or flux
- joints do not leak (particularly at bath and basin tap connections and traps, WC flush pipe connections to cisterns, and compression joints generally – which can be checked by feeling with the hand)
- pipework is adequately supported with clips as specified, at specified centres attached to firm structural support, with adequate space at back for insulation, and with falls to avoid formation of airlocks and to assist drainage
- hot water cylinders are of the specified size
- pipework, hot water storage cylinders and cold water storage cisterns are insulated as specified – with special attention to thickness, security of fixing, continuity at bends and around valves, and areas susceptible to freezing conditions (including pipes run in outside walls and unheated spaces)
- valves are present and accessible as specified. (Installations typically include separate valves to control supplies to whole buildings and to different demises, valves to control services from cold water and feed and expansion cisterns, valves immediately before all draw off points, and valves to allow radiator and other systems to be drained.)
- fire stopping and dampers are installed strictly as specified
- radiators, sanitary ware and fittings are stored and protected to ensure they are not damaged
- WC cisterns are mounted at the correct height, flush adequately and do not fill too slowly

- baths are properly seated down on cradles, do not move, and have feet fixed to spreader plates
- overflows and warning pipes are of specified size, are turned down at least 50 mm below the water line of cisterns, are laid to adequate falls, and are supported, insulated and terminated as and where specified
- radiators are fixed to slight falls to allow air to rise to bleed valve positions
- radiator plugs, valves and bleed valves are put in with PTFE tape
- radiators are securely fixed, are not dented, chipped or scratched, and are adequately protected
- guards are fitted to flue terminals less than 2 m above the ground, a balcony or a terrace
- systems are tested, flushed out, and sterilised, inhibitor added to primary circuits, and commissioned as specified. (In the absence of other requirements the contractor should be asked to run the heating system continuously for at least 48 hours, after which the programming controls should be tested. If a system has stored hot water the hot water tap should be turned on until all hot water runs out, and the time taken for it to heat up again should be checked. Temperatures from hot and cold water taps should be checked. The architect should check that turning outlets on or off does not affect the flow from other outlets – particularly showers, and taps in bathrooms and kitchens.)

7.13 Internal finishes

7.13.1 Plasterboarding

The architect should consider checking that:

- boards are stored on a flat surface in a dry place, and are handled carefully to avoid damage. (If boards are held off the floor, bearers should be at least 100 mm wide, be laid at 400 mm maximum centres, and be level. Boards should not be stacked more than 1 m high.)
- plasterboarding is not carried out until areas to be plasterboarded are weatherproof
- backgrounds are reasonably dry and flat
- timber supporting battens, framing and noggings are of specified widths and depths
- the specified thickness and type of plasterboard is being used
- dabs are applied vertically between continuous beads of plaster at floor and ceiling level, around openings, at corners, and around mounting boxes, as specified, for one board at a time. (Boards should be temporarily wedged at

bottom until dabs have set. Dabs should be about 50 mm thick, 250 mm long and have 50–75 mm between them.)
- supports for horizontal and vertical edges of boards and joints are provided as specified
- before fixing plasterboard, supports are provided for radiators or any heavy features or fittings to be fixed to the wall after plasterboarding
- plasterboards are neatly cut
- holes for services are neatly cut out before boards are fixed
- plasterboards to be decorated are fixed with the ivory papered side (which is sized) outwards; and otherwise with the grey side outwards
- plasterboards are fixed flush and aligned, with bound edges mastering external corners, with bound edges lightly butting, with exactly 3 mm between cut edges, with wall boards plumb and tightly fitting against ceiling boards, and ceiling boards level
- mechanical fixings to plasterboards are of the specified type, material, diameter and length, at 150 mm maximum centres, and 10 mm or more from bound edges and 13 mm or more from cut edges
- scrim is being used as specified to joints and internal corners.

7.13.2 Plastering

The architect should consider checking that:

- plaster is stored in a dry weatherproof area with a dry floor, with bags stacked away from walls
- temporary protection against plaster splashes to other work is provided as necessary
- backgrounds are sound, stable, reasonably dry, free from contamination and prepared as specified
- dubbing-out, for which a claim for additional payment may be made by the contractor, is measured before the work is done (or better still, the situation is avoided by measuring and including the work in the contract documents)
- plaster beads are being used and are fixed strictly as specified
- expanded metal lathing is fixed across junctions of dissimilar materials
- expanded metal lathing is fixed taut, with all strands sloping the same way in horizontal work, and downwards away from the finished face of the plaster in vertical work, with type and centres of fixings, laps, and ties at laps – all as specified
- the specified type of plaster is being used (but is not being used beyond the expiry date on the bag)
- plaster is mixed and applied using clean water, tools and equipment

- plastering is not carried out when the substrate is frozen, or during extremely cold, moist, hot or dry conditions
- each coat is of the specified thickness, and keyed and dried before applying subsequent coats
- plaster does not bridge damp-proof courses – including chemical damp-proof courses
- plaster is smooth. (Defects may not be obvious to the eye until walls are decorated and permanent lighting is switched on – by when it is too late. Plaster should be checked before decorating by running a hand over the wall or shining a torch along its surface. Attention should be paid to areas around junctions between walls and between walls and ceilings, and above doors and skirtings. The architect should look out for hollows, ripples, rough areas, flecks of plaster that have fallen from the float onto the finished plaster, and trowel marks.)
- internal angles between walls, and walls and ceilings should be straight – and vertical or horizontal as appropriate
- the finished face of the plaster is set flush with door and window frames.

7.13.3 Wall and floor tiling

Before work begins the architect should be satisfied that the contractor understands how tiling is to be set out, and how angles and edges are to be treated. The architect should check that:

- the background (including dryness), preparation, priming, adhesive and grout are strictly as specified
- setting out is as specified
- movement joints are incorporated in internal corners and where otherwise specified, are of the specified width (typically 5 mm), extend throughout the tile and bed, are continuous throughout the tiling, are free of mortar or adhesive, and are sealed as specified
- joints are straight, level and of uniform specified width, free edges are level or plumb as appropriate, and tiles are laid flat and flush at joints. (The architect should check by walking on floors and feeling walls by hand.)
- joints are fully and evenly grouted, and are slightly recessed
- edges are as specified – including threshold strips
- natural stone is grouted, cleaned and protected with sealant strictly as specified
- tiles are adequately protected after laying – with boards (laid with any ink face up) to trafficked areas.

7.13.4 Timber and sheet flooring

The architect should consider checking that:

- materials are as specified and stored correctly. (Hardwood flooring should be checked immediately upon delivery so that delays are minimised if the material is not acceptable. Quantities of knots, sapwood, and quality generally should be checked. Attention should be paid to maintaining the specified moisture content.)
- base and preparation are as specified, with particular attention to damp-proofing, levelling, flatness, smoothness, dryness and cleanliness. (Flatness should be tested with a straight edge.)
- setting out is as specified
- adhesives and fixing are as specified
- gaps to perimeters of timber flooring are as specified
- hardwood flooring is filled and sanded as specified
- sealants to timber flooring are applied strictly at the specified rate of coverage, and are not spread too thinly – to ensure the required degree of sheen is achieved
- all floors are covered up and adequately protected after laying – with boards (laid with any ink face-up) to trafficked areas.

7.14 Built-in furniture and fittings

In special cases, such as where high-specification one-off joinery is being made, it may be appropriate for the architect to visit the workshop at one or more stages during construction.

To minimise delays if replacements are needed the architect should check for damage as soon as items are delivered.

The architect should consider checking that:

- the contractor has taken measurements correctly and not, for example, assumed that rooms of an existing building are necessarily square
- items are not brought onto site too long before they are needed, and are properly stored, covered and protected
- materials are as specified
- workmanship is as specified – particularly at joints and scribings
- fixings are of the specified type, material, and size, and are fixed at all specified locations
- the operation of all moving parts is as specified
- the quality of finishes is as specified.

7.15 Painting and decorating

7.15.1 Generally

The architect should consider checking that:

- materials in storage and in use are protected from temperatures below 5°C and above 30°C, and are not too old
- all surfaces are dry and prepared strictly as specified – including removal of old paint, filling and levelling holes and depressions with stopper or filler as appropriate, rubbing down, slight rounding of arrises, and cleaning after rubbing down. (Timber should be rubbed down with the grain. The architect should inspect surfaces immediately before painting.)
- paint is prepared as specified
- knotting is applied to knots in timber as specified
- all timber, metal, concrete and other surfaces are primed as specified, working the primer well in, with special attention to the tops and bottoms of doors and windows, to the end grain of weather boarding, and to joints, angles and other potentially vulnerable areas. (Metal should be primed immediately after preparation. Special care should be taken to ensure that priming is not too thin.)
- the number of coats is as specified. (The architect should consider asking for a wall to be painted in his presence with the specified number of coats – which can then be used as a control sample.)
- drying, light rubbing down with fine paper, and cleaning between coats are as specified
- finished work is free of protruding dirt or bits of dried paint, discoloration caused by bleeding through of underlying substance, blistering, loss of gloss caused by damp in the atmosphere, brush marks, shrinkage, crazing, tackiness or softness of surface, efflorescence, flaking, undercoat grinning through finishing coat, runs and drips, saponification, glossy patches or streaks, and shrivelling.

7.15.2 External painting

The architect should consider checking that:

- painting is not carried out in direct sunlight, driving rain, fog or frosty weather
- cracks in old render are filled with epoxy filler, or otherwise as specified
- gutter boards are painted before fixing gutters, and other timber that is to be partially covered up is painted before fixing

- sealing around doors, windows and other openings in external walls is carried out after painting.

7.15.3 Internal painting

The architect should consider checking that:

- painting does not take place until dust-generating activities have been completed
- painting does not take place when temperatures are below 5°C or when condensation is likely to occur before drying
- insides of kitchen units, fitted cupboards, meter cupboards, below-stair cupboards, behind radiators, tops of doors and other out-of-the-way places are all painted
- electrical faceplates are removed before painting.

7.15.4 Wall coverings

The architect should check that:

- wall coverings are stored out of direct sunlight
- walls are primed or sized as specified
- rolls with different shade batch numbers are not used in the same room
- lining paper is fixed at right angles to the line of the final wall covering, and allowed to dry for 24 hours before hanging wall covering
- wallpaper is fixed with the specified adhesive, is properly butted at edges with the pattern aligning and the right way up, with edges trimmed neatly, with adequate paste applied to edges, with no lumps or bubbles under the paper, and with the paper clean and free of stains and tears. (Strongly patterned papers should be hung symmetrically in relation to protrusions or interruptions such as chimney breasts and doorways. The architect should agree with the contractor the precise setting out before the work is started.)

7.16 External works

The architect should consider checking that:

- materials are as specified and are stored correctly
- retaining walls are built strictly as specified
- fence post holes are vertical and of specified plan size and depth

- fence posts, straining posts, struts, and intermediate posts are as specified
- bases to paving are well compacted to prevent levels dropping and falls being lost. (It is particularly important in small areas of paving immediately north of buildings or in basements that paving is laid and remains laid to maximum falls – as even the slightest ponding will exacerbate perceptions of dampness.)
- paving units are shuffled as necessary to avoid bands or patches of a single colour shade. (A minimum of three packs may need to be opened and worked from at once to achieve the desired blending.)
- pavings are laid strictly as specified
- all builder's rubble and rubbish is removed on completion.

7.17 Practical completion

As practical completion approaches the architect should monitor progress closely. Once it is clear that the contractor has done all his own snagging, and that there are no items of work obviously outstanding, the architect should attend site to carry out his or her final inspections. In theory, the contractor should have done his job properly and the architect will find no defects or outstanding work. However, in practice, the architect should expect to find up to a small number of items in each room. Should the architect find that he is compiling a much longer list of items he or she should immediately stop inspecting and not continue until the contractor has made adequate further progress.

The architect's inspection should be made methodically in a set sequence. It is essential that throughout the inspection the architect remains alert and un-hurried. It may therefore be necessary to carry out inspections in a number of short periods rather than one long one.

It might be logical to start at the top of the building with roofs, then to work down inside room by room, finishing with circulation spaces, then to proceed to the exterior of the building, and finally to external works. At each point the architect must work through an ordered inspection routine. For example, when in a room the architect may look first at the ceiling finishes, then ceiling lighting and other services fittings, then wall finishes, then windows, then doors, then skirtings, then wall-mounted electrical and other services fittings, then sanitary or culinary fittings, then built-in storage, and then floor finishes. When inspecting decorations and finishes the architect should assume the attitude of the occupier – closing the doors to rooms, standing, sitting and even lying where the occupier would stand, sit and lie (including in the bath and on WCs!) and looking at what the occupier would be looking at. Built-in joinery and other fittings can be looked at as exercises in themselves – checking methodically from top to bottom or from inside out.

Careful notes should be taken – and recorded such that items can be very easily identified later. Lists should be subdivided by headings relating to rooms or areas, with further subdivision as necessary – perhaps by trade – to ensure that individual lists under each heading are not too long.

The architect's client should be kept informed at each stage. Above all, the architect should ensure that payment is not made for work that has not been done, and that the architect does not issue a certificate of practical completion until the contractor has attended to all but a small number of minor items – whatever the opinion of the contractor, the quantity surveyor or the client. The architect should never promise that practical completion will be certified by a particular date.

Procedures for commissioning of services should be monitored as appropriate.

Checks during the architect's final inspections should include seeing that:

- all plant and equipment is removed from site
- all doors and windows fully and smoothly open, and all ironmongery and door closers are properly fixed, adjusted and lubricated
- the heating, hot water and any other mechanical services systems operate as specified – WCs flush, taps run, all overflows work (by filling baths, basins and sinks), plugs are fitted, and all sanitary ware is clean – with all services tested, insulated and labelled as appropriate
- radiators are secure, and there is no rubbish behind them
- manholes and gullies are clear, with the correct grilles and covers fitted
- all electrical services work, there are no missing or faulty light bulbs, no light fittings are held together with masking tape, fan overruns operate correctly, faceplates are securely fixed and level, and all testing, labelling and earth bonding has been carried out as specified
- no floorboards creak
- plaster, tiles and other applied finishes are firmly fixed
- there are no chips, marks or other defects to decorations or finishes
- there are no paint spots on windows, mirrors, worktops or floors
- fixtures and fittings are all complete and working
- signage is complete
- firefighting equipment is complete
- the building is thoroughly clean, with no rubbish in cupboards
- all temporary protection and wrappings are removed
- all keys are handed over, and fit the locks
- the building manual is strictly as specified, including all test certificates, record drawings, operation and maintenance instructions, Building Regulations completion certificates and notices of other statutory approvals.

Bibliography

L. Beaven *Clerks of Works Manual* 2 edn (RIBA Publications, London, 1984).

M. Bentley *Quality Control on Building Sites,* BRE Information Paper IP28/81 (Building Research Establishment, Watford, 1981).

R. Bonshor 'Low rise housing: design and construction' *Architect's Journal* (1980), 30 April, pp 881–885.

R.B. Bonshor and H.W Harrison *Quality in Traditional Housing* Vol 3: *An Aid to Site Inspection* (DOE, BRE, London, HMSO, 1982).

R. Bonshor and H. Harrison *Traditional housing: a BRE study of quality,* BRE Information Paper IP 18/82 (Building Research Establishment, Watford, 1982).

J. Bowyer *Small Works* Supervision 2 edn (Architectural Press, London, 1979).

A. Burns *The Legal Obligations of the Architect* (Butterworths, London, 1994).

R. Cecil 'Site inspection' *Architects' Journal* (1985), 3 April, pp 75–76.

D.L. Cornes *Design Liability in the Construction Industry* (Blackwell Scientific Publications, Oxford, 1994).

I. Freeman and M. Bentley 'Quality control on site' *Building Research and Practice* (November/December 1980), pp 368–377.

R. Green *The Architect's Guide to Running a Job* 6th edn (Architectural Press, Oxford, 2001).

Levitt Bernstein Associates *Supervisor's Guide to Rehabilitation and Conversion* (Architectural Press, London, 1978).

S. Lupton *Architect's Job Book* 7th edn (RIBA Publications, London, 2000).

A.A. MacFarlane *Architectural Supervision on Site* (Applied Science Publishers, London, 1973).

P. Mayer and P. Wornell *HAPM Workmanship Checklists* (HAPM Publications/Spon, London, 1999).

J.L. Powell and R. Stewart *Jackson and Powell on Professional Negligence* 5th edn (Sweet and Maxwell, London, 2002).

Achieving Quality on Building Sites (National Economic Development Office, London, 1987).

Architects' inspection duties under Architect's Appointment and the Standard Form of Building Contract (RIBA Practice, September 1987, pp 1–2)

Health and Safety in Construction (HSE, Sudbury, 2001).

'Site architect's guide' *Architects' Journal* (1983), 20 April, pp 69–86; 27 April, pp 47–48; 4 May, pp71–77.

Index

access
- to construction in progress 64–5, 113, 115, 125
- hatch 108
- provision of 46, 51, 67, 73
- safety 26, 63, 72, 85
- to site 31

accidents 71
adjoining structures 94
appointment 1–2, 22, 39–41, 48
- documents 1–12, 15, 26, 39–41, 47

Association of Consultant Architects 1

boundaries 61, 92
briefing 54–5
British Standards 61
building contract 1–3, 10–11, 46–9, 56, 63–4, 66, 72, 78
building control officer 72, 91, 93
building lines 92
building manual 134
Building Regulations completion certificates 134

care, degree of 16–21
certificate
- Agrément 61
- of delivery 98
- of payment 2, 3, 6, 19–20, 27–8, 41, 42, 59
- of practical completion 78, 134
- of preservative treatment 105, 121
- of testing 63, 104, 121, 123, 134

certification 2–3, 6, 11, 49, 63, 65, 103
checking
- architect's responsibility for 54, 73
- checklist(s) 59–61, 62, 85, 91
- contractor 44–5, 88, 91
- failure to perform 34
- on site 91–134

- spot checks 25, 40, 46, 59, 67, 88, 89, 104

claims 76, 92
cleaning 113, 131
clerk(s) of works 9, 48
- appointment of 3, 11–12, 18, 48, 49–50
- duties of 33–6, 47, 54, 74–5, 82–3
- Institute of Clerks of Works 50

clothing 81, 82
codes of practice 61, 66, 68
commissioning 45, 125, 134
competence 17–18, 24, 33, 34
- incompetence 34

completed work 8, 62, 64, 109, 113–14,
completion 8, 12, 65, 77–9
- date 40, 52
- delay to 69–70
- practical 56, 95, 122, 123, 133–4

Conditions of Engagement for the Appointment of an Architect (CE/95) 9–10, 12
consents 22, 61, 74
consultants 10–11, 13n, 31, 33, 42, 44, 49–50, 75–6, 93, 123
costs 42–3, 46, 53, 54, 63, 69, 70, 73, 87, 89, 103
- retention money 28, 78

damage
- architect's liability for 63
- to brickwork 113–14
- to cladding 120
- to concrete 99–101
- to construction work 65, 92
- to damp-proof membranes and insulation 125
- to materials and components 64, 96, 109, 122, 123, 126–7, 130
- to pipework 125

- to roofing   107, 115–18
- to screeds   103
- to structural steelwork   103–4
- to window frames and leaves   121–2

danger   18, 27, 31–3, 51, 71–2, 82, 85
daywork   46, 74, 99
defective work   28, 43, 73, 75–6, 84
- architect's duties   2, 64, 89
- certification of   63, 78
- discovery of   44, 68–71, 88
- recording of   77

defects
- architect's duties regarding   2, 5, 17–18, 23–5, 27–8, 33, 36, 41, 47, 66, 68–71, 75, 78, 87, 133
- recording   59, 76–7, 79
- to masonry   109
- to plastering   129

delays   39, 46, 63, 69–70, 73–4, 77, 87, 89
delegation   34, 43, 75
design and build   1, 25–6
disruption   73, 74
drawings   26, 30, 43, 53, 57–8, 59, 62, 76, 81, 104, 134

duties
- administrative
- of architect   1, 3–5, 6–7, 9–10, 13, 16–19, 27–9, 35, 39–40, 42, 44, 53–4
- of consultant   31–3, 75
- of site inspectors   54, 75
- to third parties   71

employer   2, 15, 26–7, 31, 35, 45, 56, 68–9, 72, 75, 78
extensions of time   56–7, 68, 70, 76

fees   4, 8–9, 22, 24, 41–2, 43, 54, 76
fire   34, 49, 61, 71, 104, 115, 126, 134
footwear   81
foremen   62, 66, 75, 83, 84
funders   41

health and safety   42, 56, 71–2, 85
- Health and Safety Executive   72
- health and safety plan   56, 71
hours of work   50, 61
House of Lords   20, 24, 27, 29

incompetence – see competence
Intermediate Form of Building Contract for Works of Simple Content (IFC98)   64, 69–70
injury   63
instructions, architect's   2, 10, 11, 22, 42, 63, 69, 76

Joint Contracts Tribunal (JCT)   1, 46–7, 63–4, 68, 70

legal advice   41
liability
- of architect   6, 12, 27–8, 35, 40
- of contractor   27, 47
- vicarious   12, 35–6
liquidated damages   78
loss and expense   57, 76

making good   64, 69, 88, 89
management
- function of architect   10, 12
- practice management   39
- site management   56, 66–7, 91
manuals
- building manual   134
- services manual   123
manufacturers   86, 111
materials   39, 46, 67, 73, 88
- delivery of   8, 25, 45, 73, 74
- inspection of   8, 26, 51, 63, 68–69, 77, 94–6, 99, 101–3, 106–7, 115–21, 123, 125–6, 128, 130–2
- long lead-in   45, 51, 73
- storage of   45, 62, 64–5, 91, 98, 109–10, 117, 122, 123, 132
- quality of   44–7, 53, 67

means of escape   71
methods of work   29–33
method statement   66, 71
mobilisation   45
mock-up   45–6, 109
monitoring   11, 88
 – progress   10, 45, 62, 73–4, 79
 – time   42, 54

negligence   13n, 28, 35–6
 – of inspection   17, 20, 25, 48, 52
 – professional negligence   41, 88
 – vicarious liability   12
neighbours   72
notices   77, 89, 134
notice periods   46, 52, 59

occupier(s)   69, 133
opening up   8, 46, 63–4, 68–71, 88, 89, 92
 – covering up   52, 59
operatives   4, 44, 63, 66–7, 72, 75, 77, 83–4, 87, 91, 94
over-certification – see certification
owner   15, 20–1, 41

party wall agreements   56, 61, 94
payments, duty to certify   2, 6, 41, 49
photographs   76, 81
plant   45, 67, 77, 94, 134
possession   56, 77, 78
procurement   1–2, 12, 40
production information   42, 45, 50, 51, 53–5, 57–8, 66, 73, 75, 81, 91
professional indemnity insurance   49
programme   8, 10, 18, 40, 45, 51–2, 55, 56–7, 59, 67, 71, 73, 76, 123
provisional sums   45
published guidance   86

quality control   28–9, 40, 44, 55, 66–7, 71, 87–8
quantity surveyor   46, 63, 75–6, 134

records
 – contractor's records   44–5, 67, 77, 95, 125
 – inspection records   76–7, 79
repetitive elements   46, 55, 64
reports
 – contractor's   44
 – financial   11
 – inspection   75
 – site visit   59–60
retention   2

samples   45–6, 77, 109, 110, 115
schedule of conditions   91
security   45, 49, 56, 72, 91
 – of valuables   85
sequencing   5, 32, 57, 62
services   49, 57, 61, 92, 101, 102, 105, 106, 108, 123–7, 128, 133–4
setting out   51, 53, 61, 92–3, 103, 104, 120, 124, 129, 130
Standard Form of Agreement for the Appointment of an Architect (SFA/92)   7–10, 12, 16
Standard Form of Agreement for the Appointment of an Architect (SFA/99)   10–12
site agent   32, 57–8, 66, 81–4, 86
site architect   9, 12, 47–9, 61
site inspector   47–50, 54–5, 74–5, 82, 84, 87
snagging   56, 77–8, 133
specialist work   49
specifications
 – compliance with   17, 25–6, 29, 30, 94, 96, 100
 – and design   11, 43, 51, 53, 57, 59, 62, 66, 84, 88, 94
 – engineers' specifications   99
statutory approvals   134
 – consents   22, 74
statutory undertakers   74
structural engineer   27, 35, 48, 92, 93, 94, 97, 99, 100

subcontract   45, 57
subcontractors   29, 45, 55, 66–7, 72, 73, 82, 83, 84, 123
supervision   3–6, 15–19, 21–3, 25–6, 31–2, 55
supervisors   44, 84

tape measure   81
tape recorder   76
temperature   93, 97, 127
– of air   45, 119
– maximum and minimum   67, 99, 110, 112, 118, 131, 132
temporary protection   51, 56, 91, 115, 128, 134
tenants   41

tender documents   44–6, 53, 72
testing   45, 63–4, 69–70, 88, 103, 124, 134
third parties   31, 33, 41, 69, 71
tidiness   72
tolerances   61, 91, 95, 98, 101, 103–4, 110

valuations   11, 63, 76
video tapes   76
vouchers   63, 77

weather   67, 100, 103, 106, 109, 113–14, 119, 120, 123, 131
welfare facilities   45, 67, 72, 91
workmanship   53, 54, 62, 65–6, 91, 97, 110–12, 130

Index of Cases

Alexander Corfield v David Grant   22–3, 36n
Brown and Brown v Gilbert-Scott and Payne   15–16, 18–19, 23–4, 36n
Clay v A J Crump & Sons Ltd   30–1, 37n
Clayton v Woodman and Son (Builders) Ltd   29–30, 37n
Consarc Design Ltd v Hutch Investments Ltd   16, 36n
Department of National Heritage v Steensen Varming Mulcahy & Others   13n
Florida Hotels Pty. Ltd v Mayo   20–22, 36n
George Fischer Holding Ltd v Multi Design Consultants Ltd and Davis Langdon & Everest   25–7, 37n

Jameson v Simon   19–20, 36n, 43
Lee v Bateman   34, 37n
Leicester Guardians v Trollope   34–5, 37n
London Hospital (Trustees) v T P Bennett   27, 37n
Oldschool v Gleeson Construction Ltd   31–3, 37n
Saunders and Collard v Broadstairs Local Board   33–4, 37n
Sutcliffe v Chippendale & Edmondson   15, 16, 24, 36n
The Kensington Chelsea and Westminster Area Health Authority v Wettern Composites and Others   16–18, 36n
Victoria University of Manchester v Hugh Wilson   25, 37n